the Cowboy Way

seasons of
a montana ranch

DAVID McCUMBER

AN AVON BOOK

Grateful acknowledgment is given to Michael Burton for granting permission for the use of lyrics from *Night Rider's Lament,* © Michael Burton Music; to Ian Tyson for permission to use lyrics from "Springtime" from the album *Cowboyography,* published by Slick Fork Music SOCAN, © 1986 and from "When the Rocks Begin to Roll" from the album *I Stood There Amazed,* published by Slick Fork Music SOCAN © 1984; to Wylie Gustafson for granting permission for the use of lyrics from *Room to Room,* © Wylie Gustafson; to HarperCollins Publishers, Inc., for permission to use the excerpt from "Horses" from *Full Woman, Fleshy Apple, Hot Moon: Selected Poetry of Pablo Neruda,* translation copyright © 1997 by Stephen Mitchell; to W. W. Norton & Company, Inc., for permission to excerpt from "The Lady in Kicking Horse Reservoir," © 1973 by Richard Hugo, from MAKING CERTAIN IT GOES ON: The Collected Poems of Richard Hugo by Richard Hugo.

Avon Books, Inc.
1350 Avenue of the Americas
New York, New York 10019

Interior design by Kellan Peck
ISBN: 0-380-97341-3

Library of Congress Cataloging in Publication Data:

McCumber, David.
The cowboy way : seasons of a Montana ranch / David McCumber.—1st ed.
p. cm.
"An Avon Book."
1. McCumber, David. 2. Cowboys—Montana—White Sulphur Springs Region—Biography. 3. Ranch life—Montana—White Sulphur Springs Region. 4. White Sulphur Springs Region (Mont.)—Biography.
I. Title.
F739.W44M38 1999 98-46861
978.6'612—dc21 CIP

First Bard Printing: April 1999

BARD TRADEMARK REG. U.S. PAT. OFF. AND IN OTHER COUNTRIES, MARCA REGISTRADA, HECHO EN U.S.A.

Printed in the U.S.A.

FIRST EDITION

QPM 10 9 8 7 6 5 4 3 2 1

www.avonbooks.com/bard

For Sarah,

for the boys, as always,

and for the crew

Why does he ride for his money?

Why does he rope for short pay?

He ain't getting nowhere

And he's losing his share

He must have gone crazy out there

Michael Burton, "Night Rider's Lament"

contents

PART TWO
NEW GRASS

PART THREE
HAYING, SPRAYING AND OTHER DELIGHTS

CONTENTS

Part Four
THE BEST OF ALL SEASONS

introduction • birch creek

THIS IS NOT A FAMOUS MOUNTAIN.

It's not big enough, and it's in the wrong place. You can see it from downtown White Sulphur Springs, Montana, but even from that remote vantage it doesn't stand out from its neighboring peaks, some shorter, some taller. It is not quite eight thousand feet at the summit. No wisps of smoke threaten lava. People do not make pilgrimages here, though in a way I did. No pitches or crevasses or couloirs hold the bones of those who failed in some ill-fated effort to conquer it, though death visits often enough. No sleeping bags or ski poles or stainless steel watches or amusement-park rides have been named after this mountain, and you can't take a ski lift to the top, which is a good thing, though I can remember days when I would have been pleased for a ride.

On Birch Creek Ranch this mountain is called Tucker—not even Mount Tucker, just Tucker. It is a lesser peak in an obscure range called the Big Belt Mountains, which is in turn a tiny section of that enormous upthrust of granite and dime-novel romance called the

Rocky Mountains. Still. That it is a flyspeck in the scale of bigger country—the rest of this ranch, central Montana, the flatlanders' fantasy called the West—does not alter the fact that this mountain has changed my life.

Today I do not see this mountain as I have seen mountains before—as a hiker, hunter, fisherman, poet. Instead, I am trying to see it like a cow, because I am a cowboy, and somewhere in this six thousand acres of shale rock and pine and ridge and coulee and brush-lined creek are ten cows, presumably with ten calves, along with a couple of renegade bulls. Keith Deal and Tyson Hill and I are charged with finding these hard cases and reuniting them with the fifteen hundred head that came more willingly to shorter fences. Most of the herd will winter here; some of these calves will be sold and shipped to grow elsewhere. The cows, some five months pregnant even as they nurse their near-yearlings, will be tested, sorted, doctored, and stashed away to await their new calves in fields where we can get hay to them when the snow covers last summer's grass. Which won't be long. It is early November, and the west wind is blowing first rain, then snow in our faces as we move upslope, and even so my spirit is singing like Pavarotti, or at least Hank Snow.

I am a cowboy. I learned what that means from these men, and from a few others, from this mountain, and a few others, in my year at Birch Creek. I have always been a Westerner, which means I have always thought about being a cowboy. Thinking and doing are different.

This part of the ranch is called The O'Conner, after the family who ranched here seventy years ago, and it is a name that is always spoken with an edge of respect in the voice, because whatever you set out to do on this high ground will not be easy.

I came to this mountain in February, nearly two years ago now, stunned by my own ignorance. I fed hay to these cows and tried to save the calves who started life on this mountain in blowing snow and thirty below zero. We saved quite a few and lost too many. In March I took a harrowing ride up and down this mountain with these two men, through drifted snow, rescuing a pickup with no clutch or brakes. In April I came up here proposing to fix the fence that runs up the face of this mountain, and the mountain laughed and showed me great stretches of wire still under six feet of snow. In May I came back to fence again with Willie John Bernhardt, through brush so thick two hundred yards of fence took the entire afternoon. In June I watered hayfields in the shadow of this east face, and in July and August I helped to bring in the hay these cows ate last winter. In September I fixed more fence than I thought possible on and around this mountain, preparing for the gather and for the winter. For three days in October, riding a couple of the best cow horses a man could ever put his legs around, I rousted cows from this same brush with Bill Galt and Bill Loney and learned how to work the lead from Donnie Pettit, trailing cattle down the path we are climbing now. Last November, at five degrees in a ground blizzard, I chased renegade cows on foot through dense lodgepole pine timber, and just thinking about that makes today's work seem luxuriously simple.

Sometimes, in the mornings of that year, watching the sun rise over this mountain from the window of the old blue trailer that served as a bunkhouse, I would think about the cowboys of my childhood. There was Roy Rogers, who would shoot the gun out of the bad guy's hand every week on our little black-and-white TV. The Cisco Kid and Wyatt Earp and the Lone Ranger. None of them ever saw the business end of a cow, as far as I could tell. At about the same age I read the frothy Wild West creations of Zane Grey, a New York dentist, and the more serious work of Owen Wister, a Harvard Law School graduate, both Easterners who helped define the culture of the West, and in Wister's case created the cowboy culture's most enduring hero, the Virginian. To a great degree, though, I was captivated not with the cowboy as hero so much as with the land as a perfect setting for heroism. It was the mountains that got me, really, not the cowboys.

I came to this mountain for reasons of my own: journalistic curiosity about a lifestyle glorified to the point of religion in our culture—what is the Marlboro Man if not an icon, squinting down square-jawed at the mortals below on Sunset Boulevard?—and other, darker reasons, having to do with letting go of one life to find another and taking a measure of myself in a new way. It was the final step of letting go, signing on as a gray-headed greenhorn, proposing to make my living out-of-doors, with my body as well as my brain. Twenty years behind a desk in a newsroom, or rather nine newsrooms, left me searching for more. That restlessness took me five years ago from corporate striving and urban living in California back toward the beginnings of my family, to cleaner country, to something both smaller and bigger, something rougher, less despoiled, harder to win. Which is a way of saying I quit my comfortable job, started writing for a living again, divorced, and moved to a little town in Montana, near where my grandfather started this century. This cowboying, then, completed a rather thoroughgoing midlife metamorphosis.

Some of those first bunkhouse mornings I would wonder why I was spending my forty-fourth year busting truck tires or hosing out cattle trucks or shoveling cowshit out of another man's barn. It was hard to see, through that grimy little window, just where that kind of work fit into the great legend of the West, where never is heard a discouraging word, and where cowboys tip their Stetsons and say "Howdy, ma'am." I hadn't done enough of that work yet to be trusted doing anything else, or even to realize that privation and sweat and cold and danger and shit-shoveling were all part of the price of the ticket, and the payoff came on days like this, up on the mountain, just you and the cows and the grass and, if you were lucky, a horse. I hadn't spent enough time on the mountain yet to know the profound satisfaction of working on a piece of ground long enough to know it, of seeing the animals in your care thrive and grow.

Right now I'd just like to see them, period. Today I am back for a visit, for the pleasure of being on this ground, helping with the work, but so far we've had no success at all. Today Mount Tucker looks plenty big, thank you, and is giving up exactly nothing as we scan the slopes and draws for any sign of the escapees. Keith is on a dirt bike and Tyson and I are on "Japanese quarter horses"—four-wheel all-terrain vehicles. Roy Rogers never rode one, but in the late 1990s they are a noisy and somewhat regrettable fact of ranching life. The sun has pushed through, but the snow still flurries around us, and running these blatting little two-stroke bastards flat out into the wind is cold work.

The road twists us out into a big park, about halfway up the mountain, and we can see clear to White Sulphur Springs, nearly ten miles to the east, but we can't see any cows. Keith heads around to the east for a quick look, and Tyson and I move up another quarter of a mile, to a spot where we can get a better view of the terrain just below us. A few minutes later, as Keith pulls up to rejoin us, I see it. Three-quarters of a mile below and to the west a small clearing opens next to the blazing cottonwood and red willow of Butte Creek, the grass bright gold against the darker brush. As I look out across the valley and try to picture where cows might want to be, a lumpy black shape, from this far away about half the size of a dime, moves across the clearing. Bear? Elk? No, it looked too black and too square. I wouldn't have known it last year. Hell, last year I wouldn't have seen it. But today I know I've just seen an Angus cow who had rather indiscreetly allowed her black square self to be contrasted against the gold background.

"Those fuckers are in there!" I shout, and point, and Keith and Tyson look at me like I'm nuts. Nothing is visible now. It takes me a couple of minutes just to describe the clearing well enough so that they can see it. "Tough terrain," Keith said. "I'll run down there on the bike, see if there's anything there. Guide me with the radio to just where you saw it."

Handheld radios are not precisely traditional, either, but they sure work well on a deal like this. "I don't see anything down here," Keith radios after his first pass through the clearing, and I key my own mike and reply, "As you come back through, Keith, check the creek side. That's where she went." In another two minutes we hear Keith hollering, and in a few moments we see eight pairs scrambling up the hillside on the other side of the creek, and we go to help.

In the next two hours we would find the other four head, point them all down the mountain, and button them up in the pasture I'd fenced last fall. As I close the last gate in the dark I remember the first one I opened on this ranch, and think how much has opened in my life, between them. This is the story of how it happened, in my little piece of what's left of the West.

Part One

life, death and feeding hay

chapter 1 • hiplock, hard pull, heifer

THAT FIRST MORNING, THE SUN BEAT ME TO WHITE SULPHUR Springs by the barest of margins. As I approached from the south, first light spilled over the rim of the Castle Mountains and painted the stubbled hills above the town the color of a flared match.

It was barely north of zero and the wind was blowing. For the past hour and ten minutes I had been pushing my old GMC up Highway 89, through little towns still huddled under blankets: Clyde Park, Wilsall, Ringling. The surrounding country rested easy under its own blanket of snow, oblivious to the names bestowed by recent visitors: the Crazy Woman Mountains, the Bridgers, the Shields River, the South Fork of the Smith. Then White Sulphur, the Castles and daybreak, and ahead of me as I turned west onto icy gravel, the main Smith River, the Big Belt Mountains and a job I had no idea how to do.

These days there are many Montanas. I live in Livingston, a railroad town seventy-five miles south of White Sulphur. It is still at the stage of having a charmingly split personality, with cowboy bars and art galleries and coffee houses mingling in happy profusion. Other Montana towns like Bozeman have been yupped into another time zone—say, Pacific Daylight—but this place, White Sulphur Springs, Meagher County pronounced Marr, is still firmly old Montana, Mountain time, 6:45 A.M. at the moment. Big country, open, mountains on the horizon, sagebrush and bunchgrass under snow, and not a Range Rover or a Humvee anywhere.

That thought cheered me as I mashed the brakes to avoid a ribbon of whitetail deer streaming off an alfalfa field, over a fence and across the road. Ian Tyson, the magnificent Canadian cowboy singer, bounced along with me on the tape player: *"Open up the gates, boys, let my ponies roll/I'm gonna travel on the gravel, gonna head 'er for the setting sun."* Hell, Ian's still cowboying, I thought, and he's sixty. I can do this.

Right. Ian Tyson had been doing this most of his life, and I hadn't. I had come out a month before, met rancher Bill Galt, and he'd asked me what I knew about ranch work. I don't know shit, I'd replied, but he hired me anyway, maybe because at least I was honest. Honesty did not give me much comfort this morning.

The road kept climbing into the foothills, the rise approaching each hilltop a little greater than the drop on the other side. About five miles from town it crossed the Smith, and in another four the road swept to the right, but I kept going straight, through a ranch gate and toward a big, new-looking rectangular wooden building. I knew from my previous visit: this is Birch Creek Ranch and that building is the calving shed, where the heifers, young cows bred for the first time, are taken to have their calves.

I parked and got out. Something was wrong; the shed was deserted. "We get going at seven," Bill Galt had said. It was seven and there was nobody here, and suddenly I was suffused with the same panic I had thirty-eight years earlier on my first day of school, when I got on the wrong bus and ended up miles from where I was supposed to be. Bill had not told me where to report; I just assumed the shed would be the place to go. Lesson number one: You don't know something, don't assume. Ask.

I looked up at the hillside above the road I just drove and saw a large white vehicle with what looked like a huge yellow claw on the back, tilted toward the sky. A couple of people were standing on top of something on the bed of the truck. Perhaps they could use some help. I walked in that direction, and when I got over the rise I saw that they were standing on large bales of hay, the width of the truck, stacked two bales high. They were pulling on one of the bales together, their exertion showing in frosty plumes of breath. The truck was churning slowly around in a rough circle, surrounded by a milling, mooing clump of cows and their calves. When I got closer yet I was taken aback to see that no one was driving.

I waved, and got a shout in reply: "Hi! You must be David. I'm Fletcher. Come on up here and help me. Christian can drive." Come on up. Yes. Well. I grabbed a handle on the side of the cab and swung a leg up onto the gas tank, then up onto the back bed, whacking my head in the process on the big yellow thing, which turned out to be officially known as the front tines, and unofficially, for reasons I now understood, as the headache rack.

Fletcher and Christian, the *Mutiny on the Bounty* duo as I would forever think of them, were both in their early twenties. They looked decidedly un-cowboyish. Christian, hopping down into the cab and removing the bungee cord that was serving as surrogate driver from the steering wheel, was dark, intense, long-haired, and thin. Fletcher, too, was trim, a little more muscled perhaps, with a sandy beard and a dazzling smile that seemed out of all proportion to 7:00 A.M. Monday morning and five degrees. I scrambled up on top of the bottom row of bales and stood next to him, balancing precariously as Christian gunned the truck over a bump.

"Grab onto this, we've got to roll it." Fletcher pointed to the bale in front of us and I followed his example, grabbing two strings on top of the bale. I flexed hard, giving it a strong yank without waiting to get in sync with Fletcher, and I was rewarded by the bale rocking almost imperceptibly and a shooting pain in the middle of my back.

"Easy there, we'll do it together," Fletcher said, smile unabated. "These things weigh eight hundred pounds or so apiece—more when they're frozen, like these. Ready: one, two, *three*." We rocked it like that twice, and on the third time it came over—onto my right foot. My yelp was premature; I was able to get my foot out with only a little twinge. "Watch the feet, keep 'em out of the way. I'll chop, grab the strings." Fletcher picked up a homemade-looking tomahawk-like weapon with a triangular steel blade on one side and a pointed, curved spike on the other. He attacked the four orange bale strings with the triangle side. I grabbed them and pulled as he cut. "Make sure you tie those and put them with the others." I fumbled with the strings, wondering what arcane cowboy's knot I should be using. I ended up wadding them into a ball, and when I hung them up on the hook with the other strings, they came undone immediately. I turned to help Fletcher, only to discover he had already fed the entire bale and was waiting for me to help him roll the next.

This time he noticed my trouble with the strings and said, "Here, you chop, I'll tie." The chopper felt awkward in my hand, and I kept missing the strings. Finally I used it like a saw, which took longer than it should have.

The entire load took us half an hour to feed, which I figured would have been fifteen minutes, max, if Fletcher and Christian and the bungee cord had done it by themselves. Just when we finished, a four-wheel off-road buggy came zipping up alongside us. "David, come on down here." As I clambered down, the driver wheeled his buggy about ten yards away, hopped off with amazing speed and grabbbed an unsuspecting calf by the hind legs, twisted it to the ground, put a knee on it, whipped a syringe out of the pocket of his parka, gave the shot, traded the syringe for a paint stick, put a green stripe on the calf's back, and released it. I gaped, thinking how I'd like to be able to do what he had just done and simultaneously fearing that I would be expected to do just that in the next five minutes.

"Scours," he said. "Diarrhea. See his butt?" The retreating calf's hindquarters indeed bore conclusive evidence. "What's the paint for?" I asked. "So we can tell he's been doctored." "Oh. Of course. Are you having a problem with, uh, scours?"

His jaw tightened. "Too damn much. Runny eyes, too, and we've lost some to pneumonia. This weather." His eyes scanned the sky, and he pointed toward the mountains to the west. "More snow coming. My name's Keith Deal. I'm the foreman here." We shook hands, or rather shook gloves.

"This is incredible country," I said, looking around us. "Is everything we can see part of this ranch?"

"Just about," Keith said. "Everything that way, and that." We turned in a slow circle. "It goes up over that mountain, takes all the timbered part of it. That ridge over there is the Gurwell, and the O'Conner's up there above us. Lingshire, over that way about twenty miles. You work on this ranch for a year, you won't see all of it." His eyes focused back on the middle distance, at the cows and calves scattered in this pasture. "I've got to check the rest of these. You walk up the county road there. Tyson will be along in a minute and you can go with him." He sped off on the buggy, chasing another calf.

Sure enough, when I got back to the road another big white truck carrying a load of hay was churning around the corner. When I got in, I found myself shaking hands with a muscular, crew-cut young man who could have been anywhere from eighteen to twenty-five.

"Tyson." The radio crackled. "Did you pick up David?"

"Yep, got him right here. We'll start with the yearlings."

A hundred yards down the road he turned and stopped in front of a barbed-wire-and-post gate and looked at me expectantly. I jumped down, boots crunching in the snow, and wrestled the gate to a fall.

"Leave it down," he called as he drove through, a little smile betraying his amusement at my struggle.

Tyson wound the big truck through a seemingly random series of turns as we made our way across a field infinite in its whiteness and mystery. I no longer possessed a sense of direction; the seamless gray sky had lowered over the mountaintops and all I had around me was undulation of snowy pasture and the wind. Around a corner, past some frozen willows, and suddenly we were in the midst of a bellowing mass of in-between cattle, not calves and not cows, somehow the more fetching for it. Like Munchkins or Oompa Loompas, they were smaller than you'd expect yet fully developed, and they were eerily uniform in size, perhaps three feet tall and maybe five hundred pounds. Black, red, black and white, red and white, and a very few pure white, all snorting steam and bellowing for the breakfast we were bringing them.

"We'll feed here." Tyson put the truck in first gear and did the bungee cord trick, setting the wheel so the truck would turn in a short circle, and we climbed out of the cab and back to the work. Because we were carrying a full load—ten bales, five on top, five underneath—there was no place to stand to feed the first bale except on the tine rack itself. The ledge was maybe five inches wide and it felt like fifty feet above the ground, even though it was only about twelve. Tyson saw me looking at the hay and said, "We don't roll this first one because there's no place to take it. We just slide it."

"Slide it?"

"Yeah, just pull it apart from the rest so it'll flake off okay. Grab the first two strings and pull. If you only grab one and it breaks, you're going to be airborne."

I took his advice, but as we fed the third bale I bent over trying to clear away a mass of wadded-up hay, the front end hit a dip on my side, and I was airborne anyway. I pushed myself away from the side of the truck as I fell, narrowly missing snagging myself on the cable that stretched across the side of the bed, and somehow landed feetfirst right between a couple of startled yearlings. "Well, that's once," I managed as I climbed back on. Tyson kindly withheld comment.

The hay looked startlingly green against the snow, almost fake, like AstroTurf, or a brown lawn spray-painted, the way they do in Phoenix in the summer. It looked plenty good to the steers we were feeding, and across a fence a few hundred more yearlings, these all heifers, stamped and called impatiently for their share.

Three hours and seven loads later I was a wreck. The muscles in my arms and shoulders were cramping constantly. I couldn't find a position to move to so they would stop, and that shooting pain in my back was now a fixture, from shoulder blades to beltline, grabbing as I moved in any direction, stabbing when I tried to breathe deeply. Each wrench, two or three on each bale, brought a new wave of pain. On the last few trips back to the stackyard for more hay I elected to ride back on the bed in the cold because it hurt to sit down and to get in and out of the cab to get the gates. Easier to stand dazed on the bed, jouncing down a road invisible to me under snow but fortunately imprinted in Tyson's memory, or maybe the truck's. Scramble down to get gates and back up again to wonder what lapse of sanity had uprooted me so rudely this morning from home and bed and girlfriend to truck and cold and hay and pain and a landscape so savagely beautiful I knew that like so much of the mountain West it must have claimed the hearts and bodies and bankrolls of many men and women who had come before.

"How goes the battle?" Bill Galt asked with a smile. Tyson and I had finished the feeding and had taken the bale retriever around the bend past the big gate that led to the calving shed, past the entrance to the steer pasture, up a hill and and into a driveway on the right that led to the nerve center of the ranch. Tucked into a little draw were four huge grain bins, gas pumps, a beautiful old hip-roofed barn, two mobile homes, a shed, a tractor, a semitrailer, a dump truck, a big Caterpillar with a snowplow blade on the front, a front-end loader, maybe half a dozen pickup trucks, and a large rectangular metal building. This was the shop—where I was supposed to have reported for work, it turned out—and Bill was just pulling up to get gas. Dark, tall, substantial, he moved with authority and decisiveness, and when he moved, it behooved the ranch hands around him to move quickly also. Bill Galt did not like his men standing around, and he preferred them moving at a trot.

"It's going okay, thanks," I managed, joints and muscles shouting otherwise, and he nodded briefly and strode into the shop. Tyson and I followed.

"Keith, what are you working on?" Bill asked as he went in.

"Replacing the alternator on the old maroon truck, Bill."

"Good. Where are Christian and Fletcher?"

"Cleaning the calving shed."

"Okay." Bill motioned to Tyson and me. "Get these guys going on busting the tires off the crew cab and putting new ones on, will you?" I found an air wrench, had to ask how to attach it to the compressor hose, and managed to remove the wheels. Tyson busted the tires and changed them and I put them back on.

"Goddamn Ford and their better ideas," Keith grumbled from underneath the maroon truck. "You damn near have to pull this engine to get the alternator off." Nevertheless, I noticed, Bill Galt seemed to be partial to Fords. All the pickups and two of the three big hay trucks, properly known as bale retrievers, were Fords. So was the shiny new tractor parked in front of the grain bins outside. Christian and Fletcher came in, and Keith set them to work changing the oil and servicing another truck. A hand I hadn't met yet came in a few minutes later, wordlessly pulled in one of the four-wheel all-terrain vehicles and began tinkering with the throttle. It seemed as though automechanics was going to be a significant part of this job. Great. I knew about as much about carburetors as I did cows.

Keith's wife Kelly made lunch for the crew in the living quarters at the calving shed. Today's fare was lasagna, a favorite. I met Willie John Bernhardt, the fellow who'd been working on the four-wheeler. A husky, muscular six-footer in his early twenties, he nodded my way at the introduction but didn't exude any warmth. Okay, I thought, maybe it's just his way. Then again, maybe he's not thrilled that the new hand is an aging rookie.

"Fletch, go check the drop," Keith said after lunch, "and take David with you." In early February, about six hundred bred heifers had been put in the corral west of the calving shed. Half of them had since calved; the other three hundred would do so over the next few weeks. Every half hour during the day, someone walked through the corral to check the drop—see if any were ready to calve or had done so. "You look for 'em showing any sign," Fletch said. "Hiking up their tails, lying on their sides, walking away from the rest. Sometimes you'll even see a foot or two sticking out."

"What do we do if we find one?" I asked as we began walking through the big corral, eyeing heifers' nether regions.

"We take her into the shed and put her in a jug."

"Jug?"

"That's what the stalls are called. Then we watch her for the next half an hour or so, and if she's not calving okay by herself, we pull it." We were clearly an irritation to the heifers as we checked them. They would get up grumpily and move away as we approached. Fletcher talked to them kindly as we walked: "It's okay, Mom, don't worry." It was my impulse to do the same, and I appreciated him for it. He clearly liked the cows, rather than resenting them for bringing him out into the weather.

The wind had come up again, and fully half of the heifers were huddled in a mass against the west fence, seeking shelter. I didn't blame them: I was wearing thermal underwear, quilted coveralls over my jeans, and all-weather pac boots with heavy liners, and in the wind I felt as though I had come out in T-shirt and shorts. A few stragglers were across the creek that bisected the corral; we walked across the frozen stream to check them. "A lot of your calving heifers will be over here," Fletcher said. "They tend to want to get away when they start thinking about it."

"Fletch, what about that one?" I pointed to a large black Angus against the fence. Her tail was kinked, and she held it out and away from her body, and seemed to be moving restlessly.

"Good eye. I think we ought to bring her in." We cut her out of the group and she walked willingly enough toward the back door of the shed. Once we got her into the tiny corral where the door was, Fletch opened the door and stood back in the opposite corner. "Give her some room. She'll figure it out." I joined him, and sure enough, she moved toward the door, backed away for a moment, then moved back toward the opening and stepped inside.

"Okay, run her down the alley." Fletcher stepped in after her and closed the door. I got behind her and hurried her down the narrow corridor behind the stalls. One stall had its back gate open in readiness, and when she turned in I closed it behind her. Agitated now, she spun around and around the little jug.

Keith came out and inspected her. "I'll take care of it. You guys go help Willie in the chute shed."

The chute shed, across the corrals, was a much older, more dilapidated building that housed the squeeze chute, a device used to hold cattle while they are worked on—vaccinated, examined, etc. The squeeze chute was a steel enclosure slightly longer than a cow, with a gate that lifted up on the back and a catch on the front designed to hold the animal by the neck. The person running the chute waits until the animal is herded in and sticks its head through the front of the chute, then slams the head catch closed. Then if further restraint is needed, the sides of the chute can be moved inward, literally squeezing the animal and, at least theoretically, immobilizing it.

An outdoor runway led from the corrals to the chute, and Willie had already herded the eight head he was going to work on into the runway. Now, as he ran the chute, Fletcher and I prodded the first cow until she entered. She and the next two were heifers who needed their hooves trimmed, and Willie took care of them with practiced ease. The last five were newly purchased bulls that needed vaccination and branding, and they were a different story altogether. These boys weighed maybe a ton apiece, and they hit the chute on the run. When Willie caught them in the head catch the impact was like a '55 Buick slamming into a telephone pole.

The irons were heated in a propane-fired pot. Bill Galt's primary brand, /OO, required two irons, an O and a /, and three actual applications. While Fletcher and I gave the shots, Willie clipped an area high on the bulls' ribs, did the two Os first and then added the slash above and to the left. The smell of burning hair and skin filled the little shed. Sometimes the iron would make the bull blow up in the chute again; other times it seemed as though it was hardly noticed. When Willie John freed the head catch, the bulls would walk out relatively calmly. He would raise the back gate and Fletcher and I would push the next one in.

We were working the last one when Christian came running into the shed. "C-section" was all he said and all he needed to say. "Get over there," Willie said. "I'll finish up here and be right behind you." "Over there" was back at the calving shed, this time in the operating room, which was between the living quarters and the barn. There was a head catch in there too, but that's where the similarity to the chute shed stopped. This room was scrubbed clean and well-lit, with gleaming white walls, a concrete floor, and a large, very irritated cow in the head catch, legs tied, splayed on her side on the floor. It was the black heifer Fletch and I had brought in an hour before. Keith had tried to pull the calf but it was just too big.

Bill Galt was stepping into a pair of clean blue surgical coveralls as we walked in, and I figured he must be planning to assist the vet. He wheeled to the sink and started to scrub up, zinging orders around the little room like ricocheting .22 slugs. "Tyson, get the clippers ready and check her spinal. Fletch, get the scrubs ready, and my instruments. Put a new scalpel in there for me, but don't take it out of the wrapper. Make sure there's plenty of gut, all three sizes. Keith, do you have the lidocaine?"

A new scalpel for me. Jesus, he's not assisting, he's *doing* it.

"David, get clean," Bill snapped. "It's just as easy to scrub in case you're needed. You're not going to do me any good with mud and cowshit on your hands. Tyson, goddamn it, take that shirt off. You can't go into a cow with a long-sleeved shirt, you'll kill her. You know that. Pay attention and think, damn it. How many times must I tell you things? Where's Willie?"

Keith Deal, meanwhile, was bent over the cow with the electric clippers, shaving perhaps two feet by three feet on her side. "How's the spinal?" Bill barked. "Okay, Bill." "Sure?" Bill checked himself to make sure, picking up her tail and dropping it. She showed no reaction and her tail hung limp.

"Okay. Give me the lidocaine." He turned to me. "David, you clean? Come here. Don't touch her, but point to where I stop." He took the needle and made a thin red line down the black hide, in the center of the shaved area, then started injecting the topical anesthetic every inch on down the line. He stopped to refill the syringe, and I pointed at the place he left off. After he finished, he quickly followed the same route with a scalpel, cutting one quick, long stroke through the tough hide, where a red line bloomed. He squirted more lidocaine into the incision and then cut deeper, and the wound gaped open. A spray of blood came from the bottom of the cut. "Damn it, too many bleeders. Hemostat." Tyson slapped the clamp in his hand, and in a moment the spray slowed to nearly nothing.

Bill put his hands inside the incision. "This is the crucial part. If I perforate her uterus under there she's dead, and it's like tissue paper. There." He held the uterus up gingerly so that it was partly protruding from the incision, and said, "Go ahead and cut it, Keith, right there on top." Keith took the scalpel and made a cut in the glistening white muscle, and a pair of hooves popped free. "Take it, quickly," Bill said, and Tyson lifted a very hefty black calf from the cow's belly and swung it to the side, cleaned out its mouth and laid it down. Bill called out over his shoulder: "Is the calf breathing? Is it okay?"

"Seems just fine, Bill."

"All right. Pills!" Bill bellowed, and somebody jammed a piece of plastic with four large pink pills in it in my hand. "Drop them in the uterus," Bill said, and I did so. "Antibiotics," he answered my questioning glance. "Now, watch. This uterus has to be sewn up just so. The sides of the incision have to be folded over, like this." I watched, fascinated. Not only did this man seem to be an extremely competent and meticulous surgeon, but he was actually conducting a clinic on performing cesarean sections on cows. He finished suturing the uterus briskly and moved on to the main incision, pausing only while Keith poured an antiseptic into the opening. "Mr. Fletcher, come on over here and watch this. Now I'm sewing up the first layer, which is the . . . ?" He let the question hang and looked around at his crew. "Peritoneum," Willie John supplied as he walked in the door.

"Correct, Mr. Bernhardt. The peritoneum. Notice it is very thin, so I'm taking some muscle tissue with it." He worked very quickly, and his stitching would have done credit to an Ungaro gown.

"This in-between layer I'm sewing now is the muscle and fascia. How's that calf? Bull or heifer?"

"Bull, doing fine."

"Okay, Willie John, why don't you go ahead and tag him, give him his shot, and put him in the jug. She'll be ready shortly. Spray." Christian reached over and sprayed yellow Furex antiseptic spray all up and down the incision, and Bill rethreaded his needle with heavier gut. "Now restitching the hide is considerably less delicate," Bill said, jabbing the needle rather forcefully through the thick layer at the top of the cut. He made a few stitches quickly and then said, "Mr. Fletcher, you're thinking about becoming a veterinarian. Come here and sew up this cow."

"Bill, I'm not comfortable doing that," Fletcher said. "I haven't ever done it and—"

"This is the only way you'll learn," he said. "You guys need to know this stuff. You might not be able to get me or Doc some night and then you'd have to do it by yourselves. Come on."

Fletcher, highly agitated, took the needle from Bill. "Don't hold it down there. Remember, it's double-sided and sharp as hell. You'll cut yourself. Hold it up higher."

Fletch made a nervous stitch. "That's right. Notice that the sides of the incision are everted here, the opposite of the uterus, where they are inverted." After Fletcher took a few more stitches, Bill said, "Okay, I'll finish. We need to get her back on her feet. Keith, do you have the shots ready?"

"Ten ccs of anti-inflammatory and thirty ccs of penicillin?"

"That'll be fine." The suturing finished, Bill applied a final dose of antiseptic. "What time did we start?" Bill asked. "Two thirty-five," I said. He looked at his watch. "Straight up three. Not bad, but I like to get 'em closed up in fifteen or twenty minutes. The quicker you are, the less chance of infection. An old vet told me once, 'You can either be sterile or you can be fast.' You always try to be sterile. We keep this a lot cleaner than most calving sheds. But it's easier to be fast." He and Keith and Willie John untied the cow and got her to her feet. She walked a little unsteadily into the alley, and Tyson and I pushed her up into the jug where her calf waited.

"We'll have to watch her close with that calf," Tyson said. "Sometimes they won't mother up right and she'll kick the calf or not let him suck. She seems like a good mom, but we need to watch."

"Get this cleaned up and disinfected in here," Bill said, pointing to the floor of the operating room. "We've got a lot more to do. We need to move some of those pairs out. Keith, has anyone checked the drop?" When the heifers calved, the cow and calf generally stayed in the jug for twenty-four hours before being turned out into a pen behind the barn. If there were any health problems with cow or calf, they would be left in the barn longer.

Once they were turned out, they would be watched for another day before being moved from the pen into a small nearby pasture with other pairs. That group would be checked at least twice a day; calves with problems could be doctored and watched. Then healthy pairs would be turned out into a large pasture. That's what we needed to do this afternoon.

Everybody turned out to help move the cattle: Bill, Keith, Willie, Tyson, Fletcher, Christian and me. Running to where the four-wheelers were parked, I stopped and rifled the glove compartment of my truck, found the ibuprofen bottle and ate six. I rode on the back of Willie John's four-wheeler. My back was so sore I couldn't straddle the seat but rode instead sideways across the back rack. Willie looked at me strangely; I'm really making points with this guy, I thought. Whistling and shouting, pushing up close behind them, we got the cows across a creek and headed the way we wanted. A couple of times I had to tackle calves that tried to elude us. I was feeling proud of myself for grabbing and stopping a runaway calf when Keith called out, "David, back up. You won't have to do that so much if you give them a little more room." Oh.

When we got them situated, Bill said, "David and Christian, go back and help unload mineral in the reefer."

Mineral supplement for the cows came in fifty-pound sacks, and five tons—two hundred bags—were being delivered. We went to one of two identical truck trailers permanently parked near the calving shed and opened the side door. The trailers, known as reefers, were used for storage, and the mineral was kept in this one.

My back was screaming. Every bag of mineral I lifted increased the volume, but the ibuprofen finally kicked in, and by the time we were done it felt a little better.

Christian, meanwhile, had checked the drop again, and had brought two more calving heifers in. One of them had calved uneventfully by the time we checked her; the other was having problems.

Willie John got a rope around her neck, taking a wrap with the rope around a corner post on the jug and slowly drawing her closer to it. Tyson put some lubricant on his hand and reached inside her. I watched his face as he figured out what his hand was telling him.

"She's about there. I've got a foot. Give me a chain." Willie handed him a pulling chain, about three feet long. He made a loop and held it in his hand as he reached inside her again. In a moment he pulled a little, and one hoof came free, the chain around it visible now. "Let me get the other one," he said, and repeated the procedure.

When both hooves were out, he took two steel handles from his back pocket, each shaped like the top of a shovel handle, with hooks on the end. He hooked the handles into the chain, sat down in the straw, braced his boots against the cow's haunches and started to pull. "Come on, you bitch, push," Tyson said, with slightly less affection than Fletcher had showed earlier with the heifers. Experience is a cruel teacher.

Willie took one chain and Fletch the other, spelling Tyson, but it didn't help. The calf's back legs would come halfway out, but no farther. "Goddammit, she's hiplocked," Tyson said. "David, get in here, and lift that leg up and toward her head."

I grabbed her right rear leg, and she promptly kicked me in the chest. I grabbed again, and this time managed to hold the leg up long enough to get a rope around it. "Fletch, get the puller," Willie John said.

The puller turned out to be a heavy, rather primitive-looking device that did the same thing Tyson and Willie and Fletch were doing, with a little less delicacy and more stamina. It looked at first glance like a long-handled rake, but the business end had a bar that went across the cow's haunches and chains that pulled ever tighter with the turn of a crank. I was surprised at its roughness but pleased when the inexorable steel force of it finally prevailed and the calf flopped at my feet. Willie picked it up and swung it by the heels in a half circle, clearing its lungs, then set it down again. He squirted iodine on the umbilical cord and gave it a shot and copied the number off the mother's ear tag onto a new tag and loaded it into the ear tagger. It looked like a pair of pliers, with a slot for the ear tag on one side and a metal pin on the other, and Willie handed it to me and said, "Left ear." I put it on the calf's ear and squeezed, and it drove the pin through the ear and the tag, holding it in place.

Willie walked back into the shed, picked up a spiral notebook and pen off the counter next to the sink, and carefully wrote, "Number 361, Hiplock, Hard Pull, Heifer." I noticed the previous entry on the page in the same handwriting—"Number 289, C-Section, Bull"—and realized he must have made the entry after today's surgery.

"Okay, it's six-fifteen, let's bring in the drop," Tyson said, and he and Christian and Fletcher and I went out into the pasture with all the heifers and herded them into the alley behind the shed for the night. About half the heifers were eating at the manger, a long L-shaped feeding area we kept stocked with hay along one side of the pasture. I went over there to push them toward the gate and slipped on the shit-covered ice, landing on my ass. I looked up quickly, embarrassed, but either nobody noticed or they didn't care. I dragged myself up wearily and headed inside with the others.

The lights were on in the calving shed when we went in. It was probably only thirty-five degrees or so in the barn, but the contrast to the outside temperature was significant and gratifying; it was nearly dark and the cold was clamping down again.

Frank Grigsby, the night calver, was on duty, and Keith was walking with him along the jugs, pointing here and there, talking about each animal, handing things over to him for the evening. The barn was crowded; only five of the eighteen jugs were vacant. They were laid out nine to a side, with an alley behind for moving cattle back and forth and a large open area in the middle. At either end of the shed were large sliding doors that could open to admit large trucks into the center area; a big old Ford truck with a dump bed, blue in the spots where the paint survived, was parked in the middle of the shed now, heaped with dirty straw and manure from the cleaning earlier in the day. At the other end of the center space, near the door to the operating room, were several bales of hay and straw, and a large flatbed trailer with building materials stacked on it—drywall, sheathing, paneling, roofing. I wondered what that was for.

"The C-section's right here in five," Keith was saying to Frank.

"She seems to be mothering him okay. You might see if you can get him up and sucking later.

"That calf with scours is still down in eight. He's been doctored today, so you shouldn't have to mess with him. Now, these two just calved. This one was a hiplock, but they both seem to be doing okay. You've got a red heifer out there in the lot that looks kind of suspicious—number 349, she is. She might be ready for you in an hour or so; keep an eye on her."

Frank Grigsby nodded. He was an older hand, in his late forties or early fifties, with silver hair and the bowlegged gait of a lifelong horseman. "Got it, Keith."

One jug was occupied not by cow and calf but rather by a llama, dark brown with black markings, and her baby, an impossibly skinny little conglomeration of legs and ribs and neck. As Keith and Frank walked back toward the operating room, Keith jerked his head in the direction of the llamas and said, "There's fresh milk in the refrigerator. You might try to feed that camel a bottle. He hasn't eaten for about three hours."

"What's the matter with him?" I asked. "The mother never got her milk," Keith said. "I don't know if the little guy will make it or not. He's not eating that much."

Fletch and Christian and I raked up the center area. My bones ached and my eyes crossed with fatigue. "Where should I put this?" I asked Fletch, pointing at the pile of straw and manure I had gathered.

"Right there with the rest on the back of Lucy," he said, and pointed to the old Ford truck.

"Okay, you guys, get out of here, see you in the morning," Keith said.

"Where do I go?" I asked.

"Nobody showed you the bunkhouse?" Keith said. "Go with these guys. Their place is right next door."

"There's no bed in there but mine," Frank Grigsby said. "That couch probably sleeps okay."

The wall would sleep okay tonight, I thought. Keith said, "We'll mention it to Bill tomorrow. He'll get you squared away. And help yourself to any of the meat in the freezer. Bill provides that for the hands."

I climbed into the bale retriever Fletch and Christian had parked outside the shed after feeding in the morning. It was the vehicle I had started this day in, or rather on top of. As Christian fired it up and started winding our way without benefit of lights back down to the county road, it seemed at least a week since that first hay bale. It had in fact been a little more than thirteen hours.

The bunkhouse turned out to be the elderly blue trailer near the shop. Fletch and Christian lived in the yellow trailer right behind it.

"Holler if you need anything," Fletch said.

"What time's breakfast?"

They looked at me oddly. "Nobody fixes breakfast. You're on your own. You just need to be ready for work by seven," Fletch said. "Tell you what, I'll stop by to make sure you're up, and we'll have a coffeepot on."

"Okay," I said. "Thanks."

The bunkhouse was funkadelic, very seventies. The propane stove top worked great, but the oven wouldn't light. Everything in the refrigerator, I noticed, was frozen. The dining table surface was so sticky that it made a Velcro sound when you took a dish off it. I thought it was just dirty, but the surface tack survived soap, water and steel wool.

The living room offered an orange velour sofa and love seat with football-sized holes in the upholstery, and a ratty but very comfortable La-Z-Boy. Down the hall was a washer and a dryer. The washer worked fine but the dryer was useful only as a hall table. Frank's bedroom was in back; the other was tiny and featured a clothesline rigged from closet to light fixture—Frank's answer to the dryer dysfunction—an ancient, enormous and equally unusable television set, a vacuum cleaner which I suspected worked about as well, and several sets of ancient, tattered work gloves, God knew whose.

The place actually bore little sign of previous occupants, although there were a few things: a Copenhagen Pro Rodeo sticker on the front door; a few clothes in the closet. Miss December from some year or other appraised me languidly with impossibly blue eyes from her perch on the back of the bathroom door.

Frank Grigsby would make a good roomie. For one thing, with his schedule, it looked like he'd always be gone when I was off work, and vice versa. For another, the floors and counters were sparkling clean.

I went out to the mud room that had been added on the front of the trailer and rifled the freezer. Most of the meat out there seemed to be internal organs and tougher cuts, but I found one rib steak, thawed it in the micro, pan-fried it (no broiler) and ate it with a can of corn, a slice of bread, a glass of water and another handful of Advil.

As I laid out my bedroll on the orange couch, I could no longer deny the pangs of disconnection from my normal life and family and friends. I knew rationally that this was only one day and what I was feeling was the anticipation of pain, rather than the pain itself, but it felt real nevertheless. I could tell that my life would be very isolated here. The ache that was always with me, missing my two boys who lived away from me ten months out of the year, seemed sharper tonight, just because I was so far from anything I'd ever shared with them. And I couldn't keep from thinking about my girlfriend, Sarah, whom I had left sleepy and freckled and warm and sweet-smelling early this morning in my bed. Would the separation be too much for either of us to stand? Would our relationship survive it? Several times during the day I had thought of her, how she would have enjoyed the landscape, the animals, the people. The last thing I thought before sleep was that I had so much to tell her: feeding, branding bulls, C-section, hiplock, llamas. The wind was blowing hard, and it made the old trailer creak and moan. Through the window above the couch I could see snow swirling pale and cold in the moonlight.

MARCH 14

At seven-fifteen this morning, Tyson was nosing the diesel retriever into the big corner stackyard, gingerly—the approach is a sheet of frozen mud—when three whitetail deer bounded out of the stacks and headed toward the back fence. As they leapt the wire, two of them crashed into each other and the smaller doe fell heavily between the top two strands of barbed wire. She thrashed wildly as I jumped out of the truck and started forward to help her, but before I could get there she untangled herself and fell to the ground outside the fence, twitching. Bright arterial blood poured from her thigh. She couldn't move her hind legs; I think she must have broken her back, either hitting the fence or trying to free herself.

"Damn it, she's dying," Tyson said. "All we can do is cut her throat. Do you have your pocket knife?" I did, and perhaps twenty seconds after we'd driven in she was dead, a skinny skittery little yearling doe who was startled in the midst of breakfast and jumped badly. I shook my head, cleaned my knife and clambered back over the fence. A few wisps of deer hair fluttered in the wind on the second strand and for some reason I put them in my pocket. A caddis fly to tie, perhaps, or maybe just a small handful of something to remind me how quickly a life can fall beyond the fence, how quickly bottomless brown eyes can turn cloudy in the sunshine of a bright winter morning.

chapter 2 • the last of 5,000

THAT FIRST SNOW DIDN'T AMOUNT TO MUCH, BUT DURING MY second week at Birch Creek Ranch the wind started blowing hard, and for three days it only increased in strength. In this country, when the wind blows like that in the dead of winter, steady out of the west, it is like a telephone call after midnight—the news is rarely good.

Thus far, I had barely noticed the weather. It just seemed a given—cold, wind, a little snow. I was too busy getting a handle on this new existence to think about anything else. I was sore, those days—the orange couch felt fine as I collapsed on it each night, but lumbar support was not its biggest asset. Yet I was also strangely exhilarated and anticipatory. Like starting a jigsaw puzzle and finding all the edge pieces, I was beginning to define the parameters of my life on the ranch, putting another few pieces in place each day. "We need to check oil, gas and hydraulic fluid in the retrievers first thing, warm 'em up and scrape the windshields," Fletcher told me on the second morning, "and we start the tractor for Bill so he can haul grain to the yearlings and the fats. Also, every morning Christian and I feed the hounds."

"Hounds?"

"Yeah, they use them for mountain-lion hunting. They're in a kennel behind the shop. I'll show you."

The first day had seemed like a jangled, disconnected series of jobs, most of them beyond my familiarity and close to the limits of my physical ability, giving me a dropping-off-the-edge-of-the-earth uneasiness. But despite the fact I was still operating largely in a foreign dialect ("Fats?" I muttered to myself. "What the fuck are fats?"), here were understandable tasks, feeding dogs and pulling dipsticks, and they represented the first step in making myself a part of the fabric of this place. Ritual is always reassuring, the sequence of small things that frame the edges of a life, and each day had brought a little more self-assurance: *I can do this.*

I was getting to know the retrievers pretty well, particularly the Ford diesel that Tyson and I used. Dual tires on the rear, all four chained. Big yellow-painted Caterpillar engine. I even knew what to expect when I checked its precious bodily fluids. Oil okay. Antifreeze okay. Didn't use much fuel, but I usually topped it up anyway. Power steering fluid low just about every morning; the pump leaked badly. Hydraulic fluid, slow leak.

I would watch Tyson jockey the retriever expertly over hard-packed snow and frozen mud to line the ass end up perfectly square to a haystack, then reverse to within four feet of it. Then he would bend toward me and grab a red knob on the floorboard with both hands. "I'm not getting fresh," he said the first time he did it. "This PTO is pretty sticky. Something wrong with the cable."

"What's a PTO?"

Tyson strained on the knob, which finally relented and lifted a couple of inches off the floor. A red light appeared on the dashboard. He straightened, red-faced from exertion. "Power takeoff. It engages the hydraulics. Watch this, you'll be doing it soon." Manipulating a bank of levers between the seats with practiced ease, he lifted the bed and clamped the rear forks and the headache rack onto a load of bales, lowered it into place, then slammed the sticky PTO knob down with the heel of his hand, found a gear and pulled back onto the road.

That first week, rolling each bale was another pin in the lower back from whoever was holding my personal voodoo doll. Standing on the bales gathering and tying the strings still felt like trying to play the piano in a rowboat, and I don't even play the piano. One time I let the strings drag off the bed and onto the snow, and they got caught in the tire chains on the left rear dual. I had taken a wrap around my hand with the other ends as I was trying to haul them up, and when the chains grabbed I nearly went flying off the truck again. As it was, it felt like my right shoulder was being jerked out of the socket before I shook the glove off my hand. My yell alerted Tyson, who stopped the truck and backed up slowly while I jumped down and extricated strings and glove from the wheels. "You can get killed like that," he said matter-of-factly, and I promised to make every effort not to.

Most mornings, by the time we got to the second or third load, a four-wheeler would zip up to the retriever and Willie John, finished checking calves, would climb up to help us. He'd look like something between an icicle and a goalie for the Boston Bruins, wearing a rubber face protector that kept his nose from falling off—wind chill is multiplied several times on a buggy, particularly at the speeds Willie drives. Most of the time he'd take the wheel of the retriever, and Tyson would join me on top of the load; whenever a job involved a vehicle, I'd noticed, the senior crew member present usually drove.

Those first mornings Willie spoke only when he had to, and then a handful of syllables usually sufficed—"get the gate," "flake that stuff thinner," "hurry up," "toss those strings on the pile." Okay, fine. Christian and Fletcher were happy to have another hand to help with cleaning jugs, a delightful diversion each morning after feeding. Fletch showed me the procedure:

"Open that back gate and move Mom and the calf out into the alley. Close it back up. There. Now go get a pitchfork and a rake. Okay. Pitch all this dirty straw onto the back of Lucy."

"Where?"

"Oh, sorry, the big flatbed, that's Lucy. There you go. Leave the clean straw, like that over by the corner post. Okay. Now rake the hell out of the wet stuff on the bottom, get it all loosened up, like this." He took the rake and scraped the bottom of the jug hard, like a dog scratching under his collar. Soon we both had a big pile of smelly black piss-soaked straw fragments, which we scooped out and added to Lucy's burden.

"Okay, now take that wheelbarrow and get a few flakes of straw from over there"—he pointed to a stack of hay and straw bales in the middle of the barn—"and spread it out in here. Take it with the pitchfork and shake it out, like this. Okay. Now go get some hay, and fill that water bucket. Open the gate and then go around and chase them up the alley and back in. There.

"Now do the next one."

Cleaning jugs the second day, I noticed that the C-section patient and her calf both seemed to be fine. When Bill came in, his prognosis was matter-of-fact. "She looks okay, but that doesn't mean much. Peritonitis takes about three days to kill them," he said cheerfully. Nevertheless, by the end of the week cow and calf were adjudged to be healthy and turned out with the rest.

New things to learn each day. One recurring theme was that common sense, rather than arcane cowboy knowledge, would help a guy avoid trouble. Sometimes my lessons were at the expense of other members of the crew—usually the junior members, which meant Christian and Fletcher.

One afternoon during my first week, the radio in the pulling room crackled. "Birch Creek Mobile One." That was the proper etiquette for trying to reach Bill Galt on the radio. Just about all the ranch vehicles had radios, with base units in the shop and the calving barn. This time the voice sounded like Christian.

Bill strode around the corner and picked up the mike. "Go ahead."

"Bill, we're having trouble with Lucy."

Bill Galt looked exasperated. "What kind of trouble?"

"We're at the dump. She was running real rough and she died while we were dumping the load, and now we can't get her to turn over at all."

"Bill, you want me to go see if I can help?" The voice was that of Gary Welch, who did a lot of work for Bill in the summers on a contract basis, I'd been told, mostly haying and farming Bill's grain. He happened to be in the shop working on some equipment that afternoon.

"Yeah, Gary, if you don't mind," Bill replied. "Christian, did you and Fletch check the oil before you left?"

There was a lengthy pause. "No, sir, we didn't."

Bill shook his head. "Well, you've probably ruined an engine," he said shortly and put the mike back on the hook with finality.

In a few minutes, Gary radioed back. "Bill, I got it to running but it does sound a little rough. It was pretty low on oil. It seems to be okay for now, though."

When Fletcher and Christian came back into the barn a few minutes later, Bill called out, "Mr. Moon, Mr. Fletcher, I did tell you before, didn't I, that you never start a vehicle on this ranch without checking the oil? Especially an old truck like Lucy? Don't hang your head, Christian, answer me."

"I don't remember, Bill."

Bill's eyes narrowed, and pink spots appeared high on his cheekbones, above his dark beard. "That's a bunch of bullshit! You know damned well I've told you that, more than once. I shouldn't have had to tell you, frankly, but I did, and so did Keith. I don't want to hear 'I don't remember' from you. You're damned lucky you didn't blow an engine today."

"Yes, sir."

"And don't sir me!"

"Okay, Bill."

The next afternoon Keith answered for the radio call for Bill. "What do you want, Fletch?"

"It's Lucy again, Keith. She won't turn over and it looks like there's something broken on the engine."

"What does it look like?"

"There's a place where I can see something that looks broken. I'm sure it wasn't like that when I checked the oil earlier."

"Are you at the dump?"

"Yep."

"I'll be right there," Keith said resignedly.

"Something that looks broken" turned out to be a hole the size of an ostrich egg in her engine casing, where a connecting rod had punched through. Bill wasn't pleased, certainly, but there was no repeat of the explosion. "You guys sure didn't help matters when you ran it low on oil the other day, did you, Mr. Moon, Mr. Fletcher?" was all he said. Perhaps he could tell they already felt awful. Lesson learned.

As the wind howled through the week, the weather became impossible to ignore. On the morning of the fourth day I discovered that fine, dry snow had ridden the screaming gusts against the bunkhouse door in the night until it was knee-high. Where it wasn't drifted it was about a foot deep, and still coming, blowing sideways, and now the wind was charging a higher price: The thermometer on the shop wall said twenty-two degrees below zero, Fahrenheit, and I didn't care to do the math on the wind chill.

Bucking the wind back to the retriever with a quart of C-3 hydraulic fluid, I tripped on a deep-frozen wheel rut in the yard. At these temps frozen snow is like rock, which is what the edge of the rut felt like when I whacked my knee on it.

Feeding was a bitch. Many of the bales were frozen together, making them nearly impossible to roll, and blizzard conditions made the back of the moving retriever like the deck of an icebreaker. I took a heavy pry bar and whacked away at frozen Birch Creek to make a watering hole in the bull pasture, and again in the heifer pasture. A little hay, a little water. A hole in the howl of the blizzard, to make survival possible.

The simplest tasks, like kicking out pairs and checking the drop, took on new difficulty. Most of the heifers were huddled against the west fence, but a few were across the creek, and in the whiteout there was no way to check them from more than a couple of feet away. Around noon we found one of the new calves in the pasture behind the barn, down and damned near frozen to death. I picked him up and carried him into the shed, laboring against the gale. "Put him in the hot box," Tyson said. It was a big gray hinged box like an enormous airline pet carrier, with a heating element to warm calves up fast. He barely struggled as I slid him in the box as gently as I could, the heaving of his lungs the only detectable movement. Tyson put a finger in the calf's mouth. "Good way to check. If the inside of the mouth is really cold, it means they're chilled way down, internal organs almost frozen. Sometimes they come back fine, though." After lunch Keith took him out of the hot box and put him on a big vinyl heating pad. "He's not doing too bad," Keith said. "Grab a bottle and see if he'll eat some lunch."

I managed to get two pints of milk into him, and that perked him up considerably. I looked in on him occasionally that afternoon as I tended to chores around the calving barn. I was preparing another bottle at about four o'clock when he started bawling, stuck out his tongue and died. "Aw, *shit!*" I yelled, but he was beyond hearing. I couldn't believe it. I thought we had managed to save a life. Keith came in from checking cows a few minutes later and shook his head. "Nothing you could have done. Put him out behind the barn and we'll take him to the dump on the next run. And clean off that mat. Make sure you use plenty of disinfectant. We don't need any bugs getting a foothold in here."

As the light began to fade Tyson hooked a load of straw bales with the retriever and Tyson and I flaked them out along the brush line in the field with the new pairs, so they would have a little better bedding. The snow and wind had not let up at all. Just as we finished, I got a glimpse of a red heifer who didn't look right. She was swinging her head from side to side, and I thought I saw something protruding from her nose. I yelled at Tyson and pointed. He hustled around and took a look. "She's got a face full of porcupine quills," he shouted. "We've got to get her in."

Staggering through drifts, yelling and whistling, we managed to get her cut out. She protested loudly when we took her away from her calf, and I could hear him bawling beneath the wind as I got her through the last gate and into the alley behind the shed. She was doubtless in a lot of pain and worried about her calf, and it was no picnic getting her inside and into the head catch. Once we got her there we could take a good look, and it wasn't pretty. She had quills all over her nose, in her nostrils, and in her mouth. "Pretty bad one," Tyson said.

"So this happens often?" I asked as I yanked out quills with the pliers from my Leatherman tool. "Yep," he said. "The cows see the porcupines as a threat to their calves and so they butt at 'em and get smacked." The removal of the barbed quills had to be excruciating for the cow. She'd had plenty of stress in the past few days, I thought. Calving, the porcupine, the storm, separation from her calf, and now this. It was difficult to extract the quills from her nose and mouth; she swung her head violently from side to side, trying to get at us, bellowing in a low angry tone that I recognized as a reaction to pain. Finally it was done. The return trip went much more quickly; as we herded her back out into the alley, then into Pen One, then out the back gate into the pasture, we could hear her calf and so could she.

Storm, Day Two. The snow had lessened a bit in the night, but it was still coming, and the cold had not broken. Tyson and I started in the tank field, calling the cows as we drove around the side of a hill, trying to find a sheltering spot to feed and above all trying not to get stuck. "Come, cows," we yelled hoarsely into the wind. "Come, *cows*." These were the first-time mothers and their calves who had just graduated from the pasture behind the barn where we'd put the straw yesterday. They came to us, most of them, but we did find a calf with a frozen foot and another down and chilled. We picked that one up and put him in the retriever, and I radioed Keith, who was checking the calves up on the Gurwell. "Bring him on in and get him warm," he said. "I'll doctor the frozen foot on my way over to the O'Conner."

When we got back to the shed, Willie John had already plugged in the hot box. Bill was there when Tyson and I carried the calf in, and he checked the calf's mouth the way Tyson had showed me. "Boy, he's cold," he said, shaking his head. He looked at Tyson. "You fed the O'Conner yet?"

"No, we're going up there now."

"You taking the four-wheel drive retriever?"

"Absolutely. I don't think we'd get up there in either of the others."

"Okay. Don't get stuck. Give me a call on the radio and let me know how they're doing."

"Birch Creek Mobile One."

"Go ahead, Tyson."

"Bill, we found four dead calves up here so far. Two of them were in the creek. It looks like the cows just had 'em right there in the creek and they never had a chance."

"Every year," Bill said wearily. "Okay. Is Keith up there?"

"Yes, I am, Bill. I'm doctoring while these guys are feeding, and I've got another one that's chilled down in my pickup."

The drifts were enormous all along the creeks and up the draws. About four hundred cows, a quarter of them already calved, huddled against the O'Conner's white hillsides wherever they could dodge a little of the wind. A mother stood over the body of a frozen calf, lowing balefully, licking the calf and staring at us. "This is a calf-killing son of a bitch," Keith said morosely, looking at her out the window of his truck. "When it warms up and then gets cold again like this, they get wet, and then they get pneumonia, and then they freeze, and then they die."

That night I listened to the bunkhouse groan in protest as the wind tried to find a way inside and thought about Montana winters. They had plagued cowmen for a hundred and twenty years, ever since overstocked Texas ranchers had cast hungry eyes toward the sea of grass on the northern plains. This was a particularly bad storm because of its severity and timing, coming as it did in the heart of calving, but I knew the country had seen far worse.

Take, for instance, the infamous "Hard Winter" of 1886–87, which changed the way the cattle business worked forever.

In the previous decade, Texas ranchers, moneyed Easterners, titled Europeans and others had flocked to Montana and the Dakotas, seduced by claims that the abundant grasses on the plains would produce fabulous wealth in the ranching business. The Mandan (North Dakota) *Pioneer*, with a boosterism common to the era, opined in July 1883 that "the man who wants to be cattle king in western Dakota [should] come at once and select a desirable location, and we venture to say that in ten years his money will load a box car."

Pioneer stockman and historian Granville Stuart estimated that by the fall of 1886, more than one million head of cattle were crowded onto the Montana range. The beef bonanza, even then, was already on the wane; winter losses had been heavier than expected for the past three years, because of cold weather and increased grazing, and by 1885 beef prices had begun to drop due to the overstocked condition of the range.

The summers of 1885 and 1886, too, were hot and dry, particularly that of 1886. In his masterpiece *Montana: High, Wide and Handsome*, Joseph Kinsey Howard wrote: "The grass began to die in July and all but the largest streams and water holes dried up. Water in the creeks became so alkaline that cattle refused to drink it. Cinders, ashes and hot alkali dust covered the range and even the furniture in the ranch houses." In September 1886 *The Rocky Mountain Husbandman* whistled past the graveyard:

> *If we could close our eyes to the fact that there are more stock on the range than ever before and that grass in many localities has been very closely grazed, the outlook would indeed be flattering. But these are stubborn facts that cannot be regarded lightly, although all may yet be well. We shall continue to have faith in a light winter.*

In a thoroughgoing historical article titled "The Hard Winter and the Range Cattle Business," Ray H. Mattison wrote:

> *As autumn passed everything seemed to presage a very severe winter. The usual fall rains did not come. The old-timers noted that what few beaver were left in the country piled up abnormal quantities of saplings for winter food; also that the bark on the younger cottonwood trees was of an unusual thickness and toughness; the native birds such as the waxwings and snowbirds*

> *had bunched together earlier than they normally did and showed uneasiness throughout the fall. The ducks and geese flew south earlier than usual. Even the range cattle took on a longer and shaggier coat of hair. The muskrats along the creeks built their houses twice their ordinary height and their fur was longer and heavier than usual. The Arctic owls, which came only in severe winters, were about.*

One of the vagaries of Montana weather little known to outlanders is that it often stays mild until Christmas. Not so in 1886. By November, blizzards of staggering proportion were sweeping the plains, one after another. In the last week of the month, unprecedented drifts and frigid temperatures were reported across the state.

After a short respite in mid-December, the temperatures turned very cold once again. The week between Christmas and New Year's saw temperatures from thirty to forty-five below zero across Montana and the Dakotas, and the snow kept on coming. Stockmen were not feeding cattle then; on the open range, the idea was that the animals would rustle up feed for themselves—an impossibility with anywhere from two to six feet of snow on the ground. Short spells of warmer weather followed by more freezing temperatures made it even worse, encasing what grass remained in thick sheets of ice.

The winter-weary ranchers and their charges then suffered through a nightmarish February. As Mattison wrote, "The minimum temperatures at Glendive from February 1 through 12 averaged twenty-seven and a half degrees below zero. At Bismarck, February started with forty-three degrees below; on the 3rd, the thermometer recorded minus thirty-four; and on the 12th, it was forty-three below again."

After the chinook winds melted the snow in March, stockmen began to discover the extent of the disaster. Losses ranged from 40 percent to 95 percent; estimates abound, but the consensus is that more than 400,000 head of cattle died on the frozen plains.

A young cowboy named Charlie Russell spent that winter on a cattle outfit in the Judith Basin, about a hundred miles from Birch Creek. When one of the ranch owners wrote to inquire about the fate of his herd, Russell painted a quick watercolor sketch to send with the foreman's reply. The picture showed an old cow Charlie could see from the window of the ranch house, emaciated to the point of starvation, hunched against the cold. He called it *Waiting for a Chinook.* It was informally titled in later years *The Last of 5,000,* which was close to the truth. The now-famous drawing, the first well-known image by the man whose work came to symbolize Montana and the West, was purchased by Wallis Huidekoper, of a pioneer ranching family, and later given to the Montana Stockgrowers Association.

The huge losses were not simply a result of the weather. Huidekoper himself wrote: "I do not believe the 'Hard Winter' was much more severe than several others, but stockmen were improvident in winter feeding."

The ranchers had learned a hard lesson. Cattle, particularly cow-calf herds, could not reliably be left to fend for themselves through the winter. A new kind of operation started up, based on smaller herds, defined range and winter feeding operations. That meant cultivating hay, and as Howard reported, acres under cultivation for hay increased in Montana from 56,000 in 1880 to 712,000 in 1900. It was inevitable that the grass, which had supported free-ranging bison by the million for centuries, would be depleted by the overstocking of cattle, but the Hard Winter served as a sort of natural correction. Numbers were reduced and the heavy snows produced rich, plentiful grasses in 1888.

Other disastrous winters would follow—in 1906 it was so cold that tracks on the Northern Pacific and newly built Great Northern lines snapped—and the closing of the range brought new nightmares of cattle stacking up against fences in drifts and perishing. But in 1906, and again with the heavy snows of 1919, the changes in range practices kept losses to about 10 percent.

Fortunately, the cow is a particularly weather-tolerant beast, partly because its fourth and largest stomach, the rumen, functions as a large fermenting vat and gives off enough heat to help maintain the animal's body heat. That works well, down to somewhere around fifteen degrees below zero. When it gets colder than that for sustained periods, all of the nourishment the cow takes in goes to generating body heat, as opposed to maintaining or increasing body weight. Under normal conditions a cow requires somewhere around fifteen pounds of grass a day; in cold weather, during pregnancy, and particularly during lactation, that number rises to as high as twenty-five or thirty pounds. For the first month or so after birth, practically everything a cow takes in is passed to the calf, so nutrition becomes doubly important.

Even now, with the advantages of technology, everything from antibiotics to hot boxes to bale retrievers, winter losses up to 10 percent of a calf crop are still not unheard of. There's no getting around it, as I was discovering. Even for the best-conditioned cattle, Montana winters are a tough proposition.

Nevertheless, every dead calf seemed like a defeat, a repudiation of the way we were doing our jobs, thus felt keenly by the crew. For Bill Galt, the losses had to be particularly agonizing, since they represented financial as well as philosophical or emotional setbacks. For my part, I felt a responsibility not only to Bill, who had entrusted me with a job, but also to the animals, who were totally dependent upon us. We spent a few hours here or there out in the weather; they were out there all the time, and their ability to survive depended on us: on what shelter we provided from the snow, from the cold, from predators; and on the food, water, minerals and medication we provided. If it wasn't comfortable, standing on top of the haystack at the manger, forking flakes into the trough below with snow blowing in my face, I imagined how it would be for those hugely pregnant heifers waiting in the frozen mud if I didn't do it.

That sense was particularly strong in relation to the new calves, who hadn't a choice about being born in a snowbank during a blizzard. It was up to us to see that they survived it. And by the third day of twenty-below temperatures, we had failures aplenty.

With Lucy incapacitated, we loaded the dirty straw from the jugs into the bed of an old maroon Ford pickup. It was a logistical challenge because there was a hell of a lot of dirty straw. Every few jugs one of us would clamber up on top of the pile and jump up and down like a clumsy white guy trying to dance to rhythm and blues, which came naturally to most of us. That compressed the straw and enabled us to pile more on. It also made us smell pretty. The last day of the storm I slung ten dead calves on top of the straw and headed for the dump.

"Have you been up there before?" Fletch asked me.

"No."

"It's easy. You go up to the tank field, like you're headed to the O'Conner, and you'll see a road that cuts off to the left. That's it. Pretty soon you'll see the birds."

Birds? I thought as I lumbered out of the barn.

I was soon grateful for the weight in the truck bed. The road, or what I hoped was the road, was just an indentation in the snow that showed where a pair of wheel tracks might be. The foot-and-a-half snowfall made for some big drifts to bust through, and I'd already been warned that the four-wheel drive was out on the truck, and so I wanted to keep the revs up. That was a little difficult because the truck's gearshift had the unpleasant habit of springing out of third gear suddenly and violently. The first time it got me across the wrist with a crack. After that I was careful to stay out of range. It was pretty drafty in the truck, not to say breezy, and I looked over to see if the passenger-side window had been left open. It hadn't, but there was a large hole in the back of the extended cab on the passenger side. I'd seen one like it in another truck years ago, and I was pretty sure this one had been caused the same way: a shotgun in the gun rack discharging. As bad as the suspension was, I wasn't surprised.

I couldn't see much up ahead; it was mostly uphill and very white. But soon, sure enough, I saw some big black birds—ravens? buzzards?—wheeling on the horizon. Then I topped the crest of the hill and gaped at a Western vista you'll never see in a Coors commercial.

Flesh-eating birds of several varieties—including a couple of bald eagles—rose in an enormous cloud, screeching protests at my interruption. Below them were maybe fifty huge heaps of straw, dusted with snow and interspersed with the carcasses of cows and calves in varying states of decomposition. Some were little more than hide and bone; others looked as though they had just stopped for a rest—except that the birds had without exception pecked out the eyes. All this was set against the picture-postcard backdrop of the Crazy Mountains. A light, fine snow was falling.

I backed the truck next to one of the heaps. Lucy was a dump truck, which made offloading as simple as the twist of a knob, but not this truck. It had to be unloaded the old-fashioned way, with a pitchfork. Something about it got me just then, as I hunched like a monkey against the weather on top of the moldering pile—the way the wind brought the taste of rot into my throat, the weight of the filthy straw, the helplessness of watching these good calves die, something. *"God damn this storm!"* I screamed into the relentless wind, tears I didn't know I had freezing to my face. The only requiem for these blighted lives was my curse and the flap and caw of the carrion birds as they waited for another pile of death and detritus to return, forkload by stinking forkload, to the frozen earth.

MARCH 27

Four more dead calves today. The weather has cleared, but it is still very cold. I finished the day's work shortly after seven, exhausted and a little depressed.

I had to go into White Sulphur for groceries tonight, so after washing my work clothes and hanging the coveralls in front of the space heater—not the safest thing, but the only way they'd dry by morning—I stepped outside into a miracle.

Living in Montana, I am used to wonderful starlit nights, but up here there is next to none of what astronomers call "light pollution" from civilization and on clear nights like this one the views are stunning. Tonight provided something very special: Superimposed on the brilliance of the northern sky, next to the Little Dipper, was a broad swash of blue-white fire, a curved trail that dwarfed the stars, as if Van Gogh had added a palette-knife daub of titanium white as an afterthought across Starry Night.

In the silly way we have of categorizing things vast and undefinable, this wonderment was known as Comet C/1996 B2 (Hyakutake), the closest comet to Earth in more than a decade. For me, it was one of the most uplifting, soul-satisfying sights possible, timely at that.

Often at night
when the heavens are bright
with the light from the glittering stars
have I stood there amazed
and asked as I gazed
if their glory exceeds that of ours.

—"HOME ON THE RANGE"

chapter 3 • a million acres of montana

THIS PIECE OF GROUND FAIRLY SHOUTS WITH HISTORY. Running through its hills and coulees, along the streams and in the wind on the grass are the stories of many men and women, famous and humble, successful and sold out, anonymous and famous. In this century alone, such a tiny moment in the way a piece of land measures time, it has been caught up in the grand plans of three of Montana's most renowned families. It has grown hay and fed cattle that bore the brands of men with brothers and sisters world-famous for their courage, of men famous in their own right. Happily, it is largely unchanged. Where once were bison, now are cattle and elk and deer and moose and bear and mountain lion. Trout still feed on grasshoppers and eagles still feed on trout. It is, in short, paradise, albeit, as I had already seen, iced paradise, paradise with a bite. But gentler country, after all, would have been more roughly treated; I often thought, these frozen mornings, how fortunate I was even to lay eyes on this ranch. There is a purity about this relatively unvisited land, as though its soul, like some aboriginal thing, has been little compromised by the appraising eyes of civilized man.

Not surprisingly, this magnificent place was much-used and revered for hundreds of years by Native Americans of many tribes. Bison were indeed plentiful; the bluff behind Keith Deal's place was used as a *pishkun,* or buffalo jump, where hunters would stampede bison over the cliff. The Crow and Blackfeet both had traditional hunting grounds in what is now Meagher County, and often clashed over them. Sioux, Piegan, Flathead, Nez Percé, Bannack and Gros Ventre all traveled through this country. The hot springs near the North Fork of the Smith River, at what is now the town of White Sulphur Springs, were sacred to them.

Lewis and Clark passed very near in 1805. In the following half-century the U.S. government rather arrogantly ceded this ground, along with millions of other acres of what is now Montana, to the Blackfeet and Crow, who thought it was already theirs. These early treaties all were broken within a few years, after the discovery of gold and other minerals.

Immediately after the Civil War, first prospectors and then settlers began to arrive. Fort Logan and Camp Baker, both between Birch Creek and Lingshire, were built to protect the newcomers from the natives, whose claims to the area were suddenly no longer recognized.

Among the early settlers and homesteaders on land that is now Birch Creek Ranch were David Folsom, who explored the upper Yellowstone with a government surveying party in 1869; Salathiel Gurwell, who homesteaded the area between Birch Creek and the Smith River that ranch hands still refer to by his name; the Mayns, the Husseys, the Jacksons, and a host of others.

Inevitably, homesteads began to be bought up and consolidated. By the early 1920s much of what is now Birch Creek Ranch was part of the vast Meagher County holdings of the Ringling family, of Ringling Brothers Circus fame. One of the brothers, John Ringling, was instrumental in building the railroad from White Sulphur Springs to the town of Leader, which gratefully changed its name to Ringling.

First John and then his nephew Richard bought land in the area, and before long the Ringlings controlled well over 100,000 acres of Meagher County land and operated one of the largest dairies in the West, as well as the railroad and their considerable ranching operations.

Then, in the 1940s, another of Montana's most prominent men began to make his presence felt in Meagher County. He was Wellington D. Rankin, a native of Missoula, and variously attorney general, justice of the state Supreme Court and U.S. district attorney for Montana.

Rankin was the younger brother of Jeannette Rankin, who in 1916 became not only the first woman elected to the U.S. Congress, but the first woman elected to a national assembly anywhere in the world.

Jeannette Rankin was instrumental in the fight for women's suffrage on the national level, as she had been in Montana, and championed a wide variety of social issues with great fervor. But the defining moment of the freshman Republican congresswoman's career came on April 6, 1917, when Congress was asked by President Wilson to vote to declare war on Germany. Jeannette Rankin was deeply opposed to war on moral grounds.

Her brother Wellington, who had urged her to run for Congress, managed and helped to finance her campaign, supported the United States' entry into the war. He urged her to vote yes, in large part because he suspected a no would end her political career. He lobbied her strongly until shortly before the vote, but urged her at the end to "go in there and vote your conscience."

She did so. "I want to stand by my country," she said at the roll call, "but I cannot vote for war. I vote no."

Although forty-nine male members of the House of Representatives voted as she did, none took a fraction of the abuse she received as a result of her pacifist principles. The *Helena Independent* went so far as to label her "a member of the Hun army in the United States." Even many of her allies in the women's suffrage movement criticized her, believing that her vote would be a setback for the movement.

Largely through the efforts of the mining companies who were used to controlling Montana politics, she was defeated for reelection in 1918.

But in 1940, her antiwar sentiments were more attractive to many in the state, and the mining forces faced something of a quandary when she decided to run for Congress again, this time against incumbent James O'Conner. O'Conner was a relatively liberal New Deal Democrat, as distasteful as Jeannette Rankin to the corporate machine. If they attacked one candidate, the other would benefit; so Anaconda Copper and its minions gave it up as a bad job and stayed out of the race altogether. Rankin won her second term in Congress by more than 9,000 votes.

Again, the timing of history had given her a momentous role to play. When the Japanese bombed Pearl Harbor, Rankin was horrified but still incapable of voting for war. She was not naive; she understood quite clearly the global politics of the time and the political implications of her vote. This time, the House vote was 388 to 1, and her vote was met with open hostility, both from many other members of Congress and from others in the country. Some people, though, saw the courage of her act, even if they could not agree with her stance. Included among those was her brother, Wellington. Seventeen years after the vote, John F. Kennedy wrote an essay entitled "Three Women of Courage," which included her story.

In her later years, Jeannette remained an activist for women and against war. In January 1968, she led a group of women called the Jeannette Rankin Brigade in a march on the capital to protest the United States' involvement in the Vietnam War. She remained active in the peace movement until her death in 1973, at age ninety-two. Although his politics were far different—he was a conservative Republican, whereas Jeannette had been a Republican in name only—Wellington shared his sister's energy and determination, but he was not as successful as she in political campaigns. He lost two GOP U.S. Senate primaries and one general election for a Senate seat after winning the party's nomination. He was also unsuccessful in a race for governor as the Republican candidate. But he succeeded spectacularly in his business affairs.

Wellington Rankin was certainly not a typical rancher. He was Harvard- and Oxford-educated, politically and fiscally one of the most powerful men in the state. He saw the value in land and believed, quite simply and inarguably, that there would never be any more of it, and it would be in increased demand as the world changed and population increased. So he began to buy Montana land—lots of it—during the Depression years.

Although he already had substantial property in other counties, Rankin's first purchase in Meagher County did not come until the war years, when he purchased the relatively small Manuel homestead in the Lingshire country. In 1944, he increased his holdings in the county dramatically, buying what were known as the Ringling Isles—ranches at Catlin, Moss Agate and Birch Creek, the last being a significant part of the current Birch Creek Ranch—totaling more than 66,000 acres.

In 1947, former congressman O'Conner died, and Wellington purchased the O'Conner place from his widow. In 1951 he bought the 31,000-acre Lingshire ranch, which was formerly owned by the Walker Land and Livestock Company.

He added a couple of other small places near Lingshire, and then in 1954 he bought the 71 Ranch, one of the largest ranches in the state and certainly one of the most beautiful. It is mostly in Meagher County, extending from four miles west of Martinsdale far into the Crazy Mountains—nearly 80,000 deeded acres plus leased Forest Service land.

Rankin also came to own the famed Miller Ranch, up on the Canadian border, totaling nearly 300,000 acres. It was estimated that by 1960 he owned or controlled a million acres of Montana—an area larger than Rhode Island and slightly smaller than Delaware. In addition to Meagher, he owned land in Valley, Blaine, Phillips, Broadwater, Lewis and Clark, Garfield, Rosebud, Sweet Grass and Wheatland counties. In terms of acreage, he was one of the largest single landowners on the planet.

During this time Rankin continued to practice law, and in the process he met a young attorney from Lewistown who was opposing his firm in a case. Louise Replogle had already begun to make her name in the state. After graduating from the University of Montana Law School, she came home to Fergus County and promptly ran for county attorney.

She served two terms in that post, during which time she made headlines by raiding a Lewistown nightclub and charging the proprietor with operating slot machines illegally. The case went all the way to the state Supreme Court and had the ultimate effect of ridding the state of the machines. The *Daily Missoulian* ran a cartoon of young Louise Replogle running after slot machines with a hatchet, in the style of Carry Nation. But it was not a moral issue to her; it was simply the law, and she was doing her job, despite the fact that other county attorneys around the state had chosen to look the other way.

It was shortly after Louise went into private practice that she and Wellington Rankin met, as courtroom adversaries. He was impressed and invited her to join his Helena law firm, which she did; and subsequently asked her to marry him, which she also did.

The ranch properties were run out of Rankin's law office in Helena. There Louise and others in the office would keep the ranch books, figure out pay for the cowboys, and keep them and the ranches provisioned. A commissary truck—the biscuit wagon—would run from Helena around to all the ranches, bringing everything from chewing tobacco and jeans and boots and tack for the cowboys to groceries and larger items of ranch equipment. It was run like the old company store, with cowboys making purchases and the amounts being deducted from their pay.

Trucks and other machinery were few and far between. Hay was put up and fed by hand; just about everything was done by hand and on horseback, so the spreads required many more men than they do now. Finding good cowboys and keeping them was a tough chore. The state employment service would send men to Rankin; some worked out and some didn't. Critics have often contended that Rankin was on the Board of Pardons and Paroles and paroled men directly to his ranches, or that he took parolees directly from the state as a standing business arrangement. He was not on the board and did not have such an arrangement; only on a couple of occasions were men paroled directly to Rankin ranch jobs. That is not to say that a fair few of his cowboys did not have criminal records; but in that, they were no different from many ranch hands all around that country at the time.

Rankin left most of the ranching up to his foremen; he was often criticized for this and for what outsiders saw as neglect. Indeed, he was much more of a land speculator than he was a rancher; but success, particularly on the scale at which Rankin succeeded, always breeds criticism.

What sort of rancher Wellington was did not particularly concern me; I was much more interested in what sort of rancher Bill Galt was. After all, it mattered little nowadays; for when Wellington Rankin died, in June of 1966, at the age of eighty-one, he left some 600,000 acres of Montana to Louise. He had already sold the huge Miller ranch, and he had plans to sell another few properties. Those sales were carried out after his death. The rest of the land, other than a few small parcels, has stayed firmly under Louise's capable ownership.

In 1967, Louise remarried, this time to cattle buyer and ranch manager Jack Galt. Jack Galt always knew he wanted to be a cattleman—"I never had any other thought in mind," he says—and by the time he married Louise, he was an extremely well-respected one. Now, at seventy-four, he has done several other things very well indeed, but has never stopped being a cattleman.

He spent his early years in the little town of Geyser, then went to high school in Great Falls and eventually to college at Montana State. Jack learned about cattle growing up around ranches, and he learned other things. His father Errol Galt was a banker in the little town of Geyser, and when the bank failed in the Depression, as so many did, he found himself saddled with a mountain of debt. Errol went to his father—Jack's grandfather—and told him he didn't know how in the world he could pay it all back. "Don't ever come around me again if you declare bankruptcy," his father told him sternly. "But it'll take me the rest of my life to pay it all off," Errol Galt said. "You *have* the rest of your life," he heard from his father, and indeed it did not take him that long. "I still remember my father telling us when he paid the last of the debt to depositors in that little Geyser bank," Jack says. By that time his father had gone on to become president and chairman of the board of a much larger Great Falls bank.

The lesson was all about honor, and doing the right thing, and Jack Galt learned it well. So did his brother, William Wylie Galt, who earned his honor at an even higher price than his father had. He was an infantry captain in World War II. After earning two Purple Hearts and two Silver Stars in North Africa and Italy, he was near Villa Crocetta, Italy, in the summer of 1944, when his battalion was twice rebuffed in its attacks on a German position. The official account of what happened next reads in part:

> *Captain Galt volunteered, at the risk of his life, to personally lead the battalion against the objective. When the lone remaining tank destroyer refused to go forward, Captain Galt jumped on the tank destroyer and ordered it to precede the attack. Captain Galt manned the .30 caliber machine gun in the turret . . . directed fire on an enemy 77mm tank gun and destroyed it. Captain Galt stood fully exposed, firing his machine gun and tossing grenades into the enemy system of trenches despite the hail of sniper and machine gun bullets. . . . Captain Galt so maneuvered the tank destroyer that 40 of the enemy were trapped in one trench. When they refused to surrender Captain Galt pressed the trigger of the machine gun and dispatched every one of them. A few minutes later an 88mm shell struck the tank destroyer and Captain Galt fell mortally wounded.*

He was posthumously awarded the Medal of Honor by President Roosevelt.

Jack's war service was also distinguished. He quit college and enlisted after the war broke out. Because of his extensive science courses in college, the army made him a medic. "I tried to become a paratrooper to get out of being a medic," he says with a smile. All that accomplished was making him a medic with the 101st Airborne

Division, where he saw action in the Battle of the Bulge and elsewhere, earning a Bronze Star as well as the Croix de Guerre from Belgium. He returned to Montana State University as soon as the war was over.

Right out of school, Jack Galt applied for a job as a cattle buyer for Henry Wertheimer, one of the largest operators in the cattle business. Wertheimer was headquartered in South Saint Paul, Minnesota, and his Montana operations were supervised by Clarence Nelson. Nelson was not at all keen on hiring a college kid, who probably wanted too much money and didn't know anything anyway. To allay his doubts Jack Galt offered to work for a year for starting cowhand's wages, with the proviso that after that year he would either receive a salary more in keeping with his work or move on. "So they gave me some car keys and a checkbook and told me to go out and buy cattle," he remembers.

Jack Galt proved his worth, and not long after that year he took Nelson's place as Wertheimer's top hand in Montana, charged with supervising his several ranches as well as continuing to buy cattle. He worked for Wertheimer in that capacity for many years, and by the time he married Louise, he was almost uniquely qualified to help her with the management of a ranching empire.

Whatever weaknesses as a ranch manager Wellington Rankin may have had are certainly absent in Jack Galt. He is a rancher through and through, and when he took over management of the Rankin empire, certain things changed in a hurry. Some neighbors had been used to grazing and watering their cattle on Rankin ground for years. There were even some cows in their herds with Rankin brands—fences were in bad repair—and Jack and Louise moved quickly to get things squared up.

In a while, Jack pursued his own political ambitions. He served for two years in the Montana House, then ran for and won a state Senate seat, which he held for sixteen years, including two years as president of the Senate. Louise, too, remained very active in politics, as a GOP national committeewoman and delegate from Montana.

Jack had eight children from his previous marriage: Mary Ann

(Tyson and Wylie's mother); Bill, who was named for William Wylie Galt; Jackie Fay; Errol, named for his grandfather; Ben; Cathy; Harry; and John.

When Bill was born, Jack was managing the KRM ranch near Malta for Wertheimer, up on the High Line near the Canadian border. Later the family moved to a ranch Jack ran near Utica, in central Montana.

For the Galt kids, ranching wasn't a way of life, it was all the life they knew. When he was in the fourth grade, Bill Galt got a lesson in the fragility of life, and it is one he has not forgotten for a moment since. "We were at the Utica ranch. Dad was gone buying cattle. We were sitting down for dinner, and my mother said, 'Where's Buzzer?' I knew right then something was bad wrong."

Buzzer was Bill's little brother, Harry. He had been outside playing just a few moments before. Bill was right there when a neighbor found him after a two-hour search, nearly a mile down an irrigation ditch from the Galts' house. Attempts to revive him failed; little Buzzer had drowned. "I couldn't believe something so awful, so totally unexpected could happen so quickly," Bill said. "One step in the wrong direction and he was dead."

Ranches are dangerous places. Bill became known as an accomplished horseman, or horseboy, by the time he was ten. But his equine skills didn't come without pain. "What we used for transportation was horses," he remembers, but a boy who uses horses for transportation is inevitably going to want to be transporting himself at speed.

One evening in Utica when Bill was about five, he saddled up a horse himself and went out riding with friends. He didn't know how to tie a cinch knot, so his cinch was loose. They were racing across an alfalfa field toward the house when the saddle slipped, and trying to right himself he stuck a foot all the way through his stirrup. "I remember being trapped there," Bill Galt says, "going at warp speed, down between the horse's legs. That alfalfa stubble removed every bit of skin from one side of my body. Thankfully Dad was out in the yard, barbecuing, and he stepped up and grabbed the horse."

* * *

Racing rough ranch horses home in the dark. Pushing things just a little past the point of safety, for the sake of speed, of competition, of knowing inside yourself that you had been tested and measured up. That would grow within Bill Galt into the cultural hubris typical of Westerners, earned with work, pain and privation: believing that with hardheaded unflinching resolve you could accomplish more than the weaker types who could never understand the glory of the man in the arena, of trying and failing and trying and finally succeeding. For now, though, it was just racing home in the dark, with predictable result. When Bill was six, he was racing home in the dark with his best friend, Pat Seeley. They were barreling down a row of feed bunks like Affirmed and Alydar in the lane at Churchill Downs when the mare Pat was riding reached over and bit Bill's stud horse on the neck. The stud spooked and tried to jump over one of the feed bunks, but because he was running parallel to it at the time, he landed *in* it, instead, which was a horrible wreck. Bill went home and told his dad he thought his arm was broken, and Jack said, "No, it isn't. You get ready for bed." Bill went to bed, but his arm hurt too much to let him sleep, and finally about midnight Jack had to agree: Bill had indeed broken his arm.

A year or so later, racing home in the dark. Bill's horse liked the going and decided to keep running flat-out, despite all of Bill's efforts to head him. The only thing that stopped him was an irrigation ditch too big to clear. He landed in the middle of it, but Bill didn't stop. He sailed over the ditch and broke his jaw.

"I remember a horse called Sooner, a Canadian horse," Bill said. "Did the same thing to me when I was nine or ten. He had a habit of running away with me. Next day I thought, I'll fix him. I tied him to the stairs in the barn and saddled him. I made a curb strap out of barbed wire, but when I bridled him I realized it was too tight. I went to duck under his neck to go get some pliers to take it off and hit him with my shoulder. That curb strap jabbed him and he took off. With the stairs.

"I went after him and cut the curb strap off him and packed those stairs back to the barn. I hammered and nailed like a son of a bitch. As far as I know they're still standing."

By that time Bill was known for being able to straighten out a spoiled saddle horse. One of the neighbors sent him a blue roan, half Shetland pony, who wouldn't act right, and Bill rode him for a year, really loved him. One day he rode him back to the neighbor's, about ten miles, to help out with branding, and the owner, Mark Rogers, rode the horse and liked him so much that Bill had to leave him there. "I always figured they would have given me that horse," he said. "I really liked him and I was kind of heartbroken when they took him back. Nobody ever knew it, but I was."

In the summers at Utica, Bill would work on the haying crew. He started out running a conditioner, a machine that turned the hay and helped to cure it after it was mown. Then it would have to be raked into windrows and baled. The summer of his seventh-grade year, he graduated to running the baler, and a year later, a new piece of equipment called a swather arrived, which cut the hay and left it in windrows for baling—a huge savings in time and labor. Bill got to run the swather.

The year after that, Bill's father married Louise, and Bill's summers changed. "Sid Mikelson took Dad's place running that ranch in Utica," Bill remembered. "He called me that next spring and offered me $400 a month to operate the swather. That would have been big money for me. I was working on Dad's crew for $200 a month—$6.66 a day, I remember it was. We used to joke about that. I thought about Sid's offer, but I turned it down and kept working for Dad. It probably wouldn't have gone over real well at home if I'd taken it."

Working for Dad meant real cowboying, working from before dawn to well after dark, every day without a break. He started on the traveling branding crew, going from ranch to ranch with several truckloads of horses, gathering pairs and branding calves. The branding lasted all through July, starting at Avalanche Ranch, on the banks of Canyon Ferry Reservoir near Townsend, then to Catlin, Birch Creek, Lingshire, and the 71. When the branding was over, the crew was split up and sent around to various cow camps. "Sometimes I got to roam around," Bill said. "I'd take a truck with two or three horses and go wherever there was trouble, bulls loose, cows loose, whatever. I never was coached on anything, how to load anything or where to get to. But of course I didn't think I needed any coaching either. I just did it.

"When I got to this ranch at Birch Creek the crew was mostly people who couldn't get a job anywhere else, a lot of them pretty rough characters. You know, I don't know if I would have taken a kid of thirteen or fourteen and stuck him on a crew like that, miles from anywhere. But it worked out okay. I remember most of them were not very good cowboys, but they all treated me fine." Riding spoiled saddle horses for neighbors as a kid was one thing, but breaking bucking horses was a new experience for Bill, and an inescapable fact of cowboy life. He became pretty good at it quickly. He had to.

Often Jack would use him as an example, to shame the older hands into improving their performance. "Why isn't that horse being ridden?" Jack would ask. Told that the horse was too rank or had bucked somebody off, Jack would say, "Billy, go get your saddle and ride that son of a bitch."

One Sunday Jack told him to grab his saddle and go up to the corral at the 71. Louise kept a dude string there, horses she liked to ride herself, and for family and friends. They hadn't been ridden all winter and they were feeling a little fresh, so Jack wanted Bill to ride each of them to calm them down, get the humps out of their backs. "I treated that string with a little more respect after about the fourth one I got on stood me on my head," Bill remembers.

"I remember coming out to the ranch one weekend in the winter. There were two fellows there, Bill Thompson and Chuck Jackson. They were looking at a tiny little Appaloosa mare named Sissy. Bill Thompson said that horse had bucked him off, and I told him if I couldn't ride that little shit, I'd give up."

Bill saddled her up behind the corral. She was so small he couldn't get the saddle tight and the first time she bucked him and the saddle off. He resaddled her, punching another hole in the leather to get it tight, and she bucked him off again. "I got back on and she sort of bucked into the side of the barn and I managed a few laps around the corral and called her rode, but she bucked me off a lot more the next summer."

The next year he decided to try bareback riding and bought a rigging. He was practicing on Sissy, coming out of a corner of the corral with her, another hand cutting the halter rope and swinging a gate open to simulate coming out of the chute, when Jack showed up and demanded to know what he was doing.

After chewing Bill out for contesting ranch horses without permission, he said, "Did you ride her yet?"

"No, she bucked me off two or three times. I'm tired of it anyway."

"Well, you started it and you're not going to stop until you get it done." Jack ran the gate for him then, and it took seven tries, but Bill Galt finally rode her.

Bill rodeoed competitively for his last three years in high school. His sophomore year he rode only barebacks, but that year he took a week off school and went to a saddle-bronc school run by champion Shawn Davis, and from then on he competed in both bareback and saddle bronc riding, making it to the state finals in saddle bronc his senior year.

He qualified for state at the district rodeo at Townsend. He had to take a day off from work to go. He hadn't realized that his father was in attendance; Jack never said anything, but he usually went to see Bill compete, as Bill found out later. The next day driving in to work Jack said, "I see you won the bronc riding over at Townsend."

"Yep."

"I suppose that means you'll be wanting another day off to go to state then, eh?"

Bill grinned at the memory. "Dad wasn't much for praise, that was his way, but you knew what he meant."

Bill drove the biscuit wagon for a while during the summer of his senior year, ferrying supplies of all kinds around the ranches. One of the chores he was given was to drive a brand-new blue Ford two-ton truck with a dump bed to Birch Creek. It was Lucy, and knowing that made me a little more understanding of why he had been so irritated at the careless treatment of the battered old truck and its subsequent breakdown.

He was born to the work, but a lot of it must have seemed like drudgery to an eighteen-year-old. He shook his head, remembering. "By the time I was out of school I was sick enough of ranching I thought I never wanted to see another calf. It took me about six months away from it to be dying to come back."

On January 1, 1972, the day after his nineteenth birthday, Bill went into the army. They took a look at this big ranch kid from Montana, six-foot-plus, a competitive bronc rider, grew up working hard, and figured he was tough enough and dependable enough to ride herd on renegade soldiers. They were right. The army made Bill Galt an MP, sent him to noncommissioned officer school, and put him to work on AWOL apprehensions, stationed at the First Army prison at Fort Knox.

After his discharge Bill, too, went to Montana State. but he and his brother Errol were running a ranch outside of Ingomar, nearly three hundred miles east of Bozeman, at the same time, so college came hard.

He drove back and forth, being forced to miss much of his classwork, then finally quit school and moved to Ingomar full-time. It was there that Bill got into the cow business, buying fifty head of bred two-year-old heifers with the help of a $16,000 loan from the Farm Home Administration. "You have to be turned down by three banks before you can get an FHA loan," Bill said, "which wasn't hard. I didn't want to ask Dad or anybody to cosign for me."

That first venture didn't go too well. Bill had to pay for the hay and cake to winter the cows, but he was allowed to pasture them for free. It was a bad drought year, and half of the heifers came in dry. The FHA made him sell the dry cows. "I think I got eighteen cents a pound for them," Bill said. "It didn't pencil out too well."

But he managed to find someone who would trade him some bred cows, and that kept him in business. Then the following year the foreman at the 71 Ranch decided to leave, and Jack and Louise asked Errol to come manage it. Shortly after that, they asked Bill to take over the Birch Creek and Lingshire operations. Then Ben was tapped to run Catlin, Moss Agate and Battle Creek. Ben's spread involved much more farming than ranching; hay and cereal grains, not cattle, were the emphasis on those ranches, which suited Ben fine. He was much more interested in farming than ranching.

For Bill, returning to Birch Creek, where he had worked as a cowhand growing up, was an equally perfect fit. It provided the opportunity for Bill to set up a ranch the way he thought it ought to be. Certainly the potential was there; Birch Creek had superior hayfields and good pasture ground as well. Lingshire offered a big additional chunk of summer rangeland. The places were largely undeveloped; corrals and buildings were primitive and quite limited.

For the previous decade, Birch Creek, Lingshire and the O'Conner had been managed by Bill Loney, a rancher who was somewhat of a legend for hard work and handiness. Loney was born in Choteau, Montana, on the Rocky Mountain Front, where his father ranched. At age fifteen he left home and went to work for rancher Charlie Hodson near Cascade, where despite his tender years he quickly became the foreman. He kept that job for a decade, then decided to head out on his own. He moved to White Sulphur, leased almost three thousand acres of ground near the Smith River and bought a few cows.

In a year or so he approached Wellington Rankin about doing some work on Rankin's Birch Creek property, which was close to the land he was leasing. "I asked him if he wanted to trade some grass for some work," Loney said. "I told him if I could put some cows on his pasture I'd fix his fences and put some brands on his calves, but he didn't want to do that. He said he might be able to use me breaking some horses for him, but we never agreed on anything."

But a few years later, Jack Galt and Bill Loney decided to do some business. Jack and Charlie Hodson had traded cattle together; in fact they had gone together a few times and bought cattle from Wellington Rankin. Jack had gotten to know Hodson's young foreman, and so after he married Louise and began to help manage her properties, he talked to Loney about looking after the Lingshire ranch. "I went up there in the spring of 1968," Loney remembers, "and then I suggested to Jack that I could take care of the land at Birch Creek, too."

Jack was more than agreeable, and so for ten years Bill Loney ran Birch Creek, Lingshire and the O'Conner and kept his place going too. Jack let him run some cattle on Birch Creek and so he was able to expand his own herd.

"Billy had worked for me up at Lingshire—Jack sent all his kids to work for me—and so when he came back to Birch Creek I stayed and helped him out," Loney says. It amounted to a two-year apprenticeship. Bill had respect bordering on awe for Loney after working for him as a kid. "Everything I ever learned about a cow, I learned from my dad or Bill Loney," he says. After a while, Loney felt like Bill was ready to run the place by himself. "There's only room for so much bossing," he says. "Billy was doing a good job, and I felt like I was on welfare or something." At the same time, Bill Galt was very sensitive about not wanting to step on Loney's toes.

One night, at a party for the crew at the end of branding, Bill Galt and Bill Loney got to talking. "Both of us had probably had a drink or two," Bill Galt said, "and I told him, 'Look, I don't mean to be getting in your way here,' and he said, 'No, Bill, you take it. I'm tired.' " After all, Loney had bought a place of his own, up near Lingshire, and he was also running a post and pole business in addition to the land he was managing for Louise and Jack. So he was ready to turn his attention to his own spread full-time. He continued to run some cows at Birch Creek in exchange for his advice and occasional assistance, and that's pretty much the way things stand today.

Bill Galt began to expand the small herd he'd brought from Ingomar, and to modernize the operation. As we talked into the evening at the calving barn, Bill waved his arm around to indicate the corrals and the barn. "We built all this ourselves," Bill said. "The chute shed used to be the calving barn. We built the shop, the scale shed, the corrals, everything."

I watched the pride in the sweep of his hand and thought, Now I understand a little bit of what I am dealing with here at Birch Creek.

MARCH 23

Love It Now

Believe in this little chinook.
Don't worry about it breaking your heart
next week. Love it now: West of White Sulphur,
old grass glows amber where snow should be.
You drove on gravel like this years ago,
sweet with the musk of cut grain
from night pastures,
farm girl's lip catching Coors
jounced from church-keyed can at speed.
Our innocence still lives here. Eight miles
from town a bull calf is winched into the world,
bootheels braced against Angus flanks.
He'll know twenty below
before he's cut and branded,
but today he can nuzzle a tit
in the sunlight, drink from the river
too trout-sacred to mention here.
Believe in this.

chapter 4 • top hand . . .

BILL GALT OWNED TWO KENWORTH TRUCKS, AN OLDER YELLOW model he drove himself and a blue truck that was a couple of years newer. The blue truck was Keith's to drive, and although both were maintained impeccably, Keith was particularly proud of the blue one and kept it in showroom shape.

"David, take the shop vacuum and clean out the inside while I change the oil," Keith said. I did so, and then took the mats out, cleaned them and Armor Alled them and then cleaned the glass inside and out. Keith was having some difficulty replacing a thermostat on the truck, which turned out to be the fault of the parts dealer, who had sent him the wrong one. He had drained the oil from the crankcase, and while he was messing with the thermostat he had me refill it. I was impressed to discover it held eleven *gallons*.

After we finished, Keith said, "You and Tyson go hook a load of hay and follow me up to the O'Conner. We've got a job to do."

The road was slick and deeply rutted in spots, but we got through, Tyson white-knuckled behind the wheel, and picked our way around the snowdrifts in the field and got the hay fed. As we finished, we saw Keith's pickup approaching. "Park it," Keith radioed. "You're coming with me." Tyson pulled the retriever up next to the road, and Keith circled down below us, out of sight for a few minutes near the creek, then drove back to us.

Tyson and I squeezed into Keith's new Ford pickup. It was a one-ton flatbed, fitted out as a service truck, with a pair of hundred-gallon gas tanks and an electric pump on the back, plus two large toolboxes. He had a set of chains on the back tires. The cab smelled like Keith had been carrying calves in it, which of course he had.

"Did you see any problems?" Keith asked.

"No," Tyson said. "They all look healthy enough. Well, one cow was calling for a calf, but I didn't see him anywhere."

"Red and white?"

"Yep."

"I just brought her calf back up here. I took him in to warm up this morning."

"Where are we going?" Tyson asked.

"Over to Doc Schendal's. Bill bought that banged-up Ford pickup he has and we've got to bring it back."

"Does it run?"

"Nope. The clutch is out of it."

"Oh, shit. This'll be interesting."

Keith chuckled.

Just getting to the truck was interesting. Schendal's spread was adjacent to the back side of the O'Conner, but there was a lot of rough country between it and us, including a swamp. But Keith Deal was equal to the task. He is the kind of driver many men fancy they are, but really aren't. I could already see he had an athlete's reflexes and the nerve of a cat burglar. Nothing fancy, no overrevving or unneccessary braking, not one degree of surplus steering, just precisely what the situation demanded.

In fifteen minutes we were parked in front of an attractive old wood-frame house, looking at a very crumpled Ford that appeared to have been through Desert Storm. Something—a fire, a defective factory paint job, perhaps both—had caused almost all of the paint on the hood to peel off. The hood was badly dented, too, from a front collision. Both doors were mangled. The sheet metal at the hinges was curled out as though the front collision had sprung the doors and they had been readjusted with a crowbar. The driver's door looked as if it had taken an extra hit of some sort.

"Wow," Tyson said. "It's not that old, is it?"

"Nope, it's an '89," Keith said. "Doc's not exactly into maintenance."

"See if the hood will open," Keith said. It did, with a little coaxing, to reveal another surprise. "There's no battery in it," I told Keith.

"Okay. Get all his shit out of it. Put those chains and all the big stuff in the yard and the rest in that cardboard box in the cab, then put it in the house."

Keith, cleaning out the glove box, found a pair of red plastic mirrored sunglasses. "All *right*," he said. "I need a pair. How do I look?"

He looked tough—and so did the way back to Birch Creek. There was a mountain and a lot of snow between us and home. I couldn't believe Keith actually planned to tow this derelict rig, but he did. We got a towrope hooked and Tyson installed behind the wheel of the battered truck. "Brake when I brake," Keith said. "Steer. Good luck."

I shut Tyson in—the door wouldn't close from the inside—and got in with Keith.

I sensed that Keith was in a little more expansive mood than usual, and so as we made a sweeping U-turn and started back, I asked him, "How did you learn to drive like this?"

As he talked about his background, what he'd accomplished, he came into focus for me—why he was so handy, why he knew so much, why he was the foreman.

Keith Deal graduated from White Sulphur Springs High School in 1976. He graduated angry because a conflict with administrators kept him from playing football his senior year, and he was passionate about football. He probably weighed 160, max, and he played guard, where he loved to mix it up, lay waste to opposing linemen and ball carriers, and he was from all accounts a truly fearsome player despite his size.

During his high school years he worked for Herb Townsend at the Townsend Ranch, one of the largest spreads in the valley, and after he graduated he took a job at the sawmill in White Sulphur, starting at the bottom, pulling lumber off the green chain, then quickly moving up to running a 966 front-end loader, unloading trucks. He'd been around engines and machines all his life; he loved cars and motorcycles, and raced Suzukis in motocross events for a while. The sawmill took advantage of his mechanical ability and made him millwright, welding, fixing things, keeping all the machinery running. After a couple of years, before Keith Deal turned twenty-one, he was promoted to foreman of the night-shift crew.

Working at the mill, Keith met a trucker named Tom Higle from Deer Lodge, Montana. Higle noticed how handy Keith was, how he could drive, and he offered him more money to drive logging trucks than he was making running a shift at the mill. It was a good change; in a year the mill was closed down anyway and Keith didn't have to sit around wondering where he was going to find a job.

Then Keith got married and moved back to White Sulphur, working for a cement plant, driving a cement truck and also eighteen-wheelers, loaders and anything else on the place with wheels. One day in 1982, driving a truckload of cement out past the nine-mile Y south of town, Keith came over a hill to find Ben Galt's cows across the highway. There was no flagger out to warn him that there was stock on the road, and he plowed into the cows, lost control of his truck and hit a Volkswagen bus driven by a White Sulphur girl he knew. She broke her neck and died. He was found not to be at fault, but Keith decided he'd never drive an eighteen-wheeler again. Six months later, he quit the cement plant and went to work for Herb Townsend again.

He stayed a couple of years before Jack Golberg hired him away to be his top hand. Golberg ran the only major construction firm in White Sulphur, and Keith spent his days building houses and putting in water lines, sewer lines, stock tanks, drain fields and the like. It was good work and Jack Golberg will tell you today that Keith was damned good at it. But there wasn't much work in the winters, and in early 1986 Keith went out to Birch Creek and asked Bill Galt for a job, feeding hay. Jim Witt, who had been the cow boss of the 71 Ranch for a while, was on the Birch Creek crew then, and Pat Bergan and Terry Alexander, and another hand or two. Bill had met Keith a few times in town, and had heard how he was pretty handy, and so he was happy to hire him.

Within a couple of years, Keith was helping Bill run the crew. "I kind of took it on myself at first to make sure everybody stayed busy, make sure the work got done," Keith said.

Bill knew Keith wouldn't drive eighteen-wheelers, and why. So one day he took Keith along when he was delivering a load of calves to rancher Tom Lane near Livingston. When they were done unloading, Bill headed for the passenger side instead of the driver's side of the cab.

"I pedaled down, you can pedal back," he told Keith.

"I don't know, Bill," Keith replied unhappily.

"By God, Keith, you'd better get over it. If you don't do it now you'll never do it," Bill said, with his typical forthrightness.

"So I hopped up into the driver's seat and away we went," Keith said. "Damn, Tyson, watch the brake lights. You're too close." I had watched him a few hours earlier, turning wrenches on his treasured blue Kenworth, and I knew that he was not only happy he'd gotten past that terrible accident but also grateful to Bill for forcing the issue.

Bill must have cussed himself for it the next year, though. In 1989, Bill, Keith and the crew built the calving barn and the living quarters. It was a huge job, and Keith's construction experience paid big dividends. But it was really stressful, too, and finally Keith decided he'd had enough of it. "I didn't make any fuss about it, I just quit and took a job driving truck over the road for Associated Foods," Keith said.

Some things happen for a reason. Keith's delivery route with Associated took him to grocery stores all over the northern Rockies, including to tiny Eureka, in the far northwest corner of Montana. He'd been through a nasty divorce, and so when he met pert, brown-eyed Kelly, the produce manager at the grocery store in Eureka, he did more than deliver her produce. They were married in June 1993, and by that fall they had decided it would be good if Keith could get off the road and settle somewhere so they could have a stable home for themselves and for Summer and Doran, Keith's kids from his first marriage.

So that September, Bill Galt made Deal a deal he couldn't refuse. He said, "Pick a spot on the ranch and I'll put a home there for you." Along with the brand-new mobile home on the banks of Birch Creek, between the calving shed and the shop, came a new pickup to drive on the ranch and the shiny blue Kenworth, too. When you factor in the house, free of any cost for mortgage, utilities, taxes or upkeep (other than Keith's own sweat), the foreman's salary, somewhere around $20,000, stretches quite a ways. It's about twice what a green hand—like me—makes.

Still, I thought to myself, Bill was damned lucky to have him. And I figured Bill thought so, too.

I envied Tyson his part of this job, steering this ghost of a truck as Keith pulled it over the mountain. It looked like fun. I knew a serious ass-chewing could result from fucking up, so I was probably better off, but I still wished I could have tried it.

Even from my passenger's perspective, it was pretty hairy. Some of the drifts were pretty deep, and Keith had to keep moving at a pretty good clip; he needed the speed to power through. A couple of times he slewed sideways, but he managed to keep going in the right direction.

"Goddammit, Tyson, *brake*," Keith muttered again. "He's going to ride that thing right up our asshole yet."

Keith was undeniably in a pressure-filled situation, being Bill Galt's foreman. A few days earlier we had started to transform that stack of building materials in the barn into a portable cabin for use in the Lingshire country during calving. We hoisted the plywood sheets, two-by-fours, roofing and other supplies off the platform and Keith had connected an air-driven nail gun and a staple gun. By about seven-thirty in the evening we'd stapled plywood sheets for the interior walls to the framing, and we'd even got windows cut out and started on roof joists. This had not been accomplished without a certain degree of testiness. It had been a long, physically demanding day for everyone. About then Bill called and talked to Keith, who came back and said, "Everybody go home." He muttered a quiet aside to Willie John: "He wants everything done, but he just told me not to have the crew work these long days." He shook his head. "Hard to have it both ways."

There, I thought, was the essence of the foreman's dilemma. It is a classic middle-management squeeze play, with the ranch ownership on one side and the cowboys on the other. Please the boss, piss off the crew, or vice versa, or maybe please nobody. And of course work longer hours than anybody else.

Keith was famous for screaming at hands who fucked up. He could do any job on the ranch, and he expected everybody to be able to do the same; often, the crew fell far short of that mark. Still, possessed of a grin like a naughty kid's and a great sense of humor, he was great to work with when he was happy.

A good foreman is a cowboy's cowboy, sort of a warrior king; he has earned his position by working harder and better than the rest of the crew, and as such is revered and perhaps a little feared. If the cowboy is a mythic figure, the cow boss has earned a spot a little higher up the slopes of Olympus.

From the beginning of ranching in the Northern Rockies to the present, absentee owners have always been utterly dependent on their foremen. Often in those early days when things went badly, the foreman ended up with control of the ranch.

After the Hard Winter, several of the large Eastern firms simply refused to provide further financial support to their ranches, and the foremen continued the best they could. In other cases, foremen were able to buy the "foreigners" out at fire-sale prices. Henry Lingshire, top hand of the Walker Land and Livestock Company, eventually bought out his Eastern bosses and took over the ranch that is now part of Bill Galt's summer range and still bears his name.

Bill Galt is no absentee owner; he is much more engaged with the daily routine than most ranchers, so the friction points with a foreman are bound to be greater. Even warrior kings, after all, have gods to answer to. I began to understand that Keith's mood was often a reflection of Bill's; if Bill was displeased, it was a cinch that Keith would be more so. As Willie John Bernhardt put it to me, succinctly, "Shit rolls downhill around here."

When we got back in sight of the retriever I figured we had it made. "You want me to drive that back?" I said, pointing to the big white hay truck.

"No, I think Tyson can do that," Keith said, slowing to a stop. "You can steer the sled back there the rest of the way." Oh, shit. Be careful what you wish for.

The driver's door of Bill's newest old truck opened with a metal-on-metal screech, and Tyson stepped out, looking a little pale. "Go ahead and drive the retriever back, Ty," Keith said.

"With pleasure," Tyson replied. He closed the door of the pickup after me and said, "Good luck. It has almost no brakes."

"I was wondering," Keith said. "Okay, David, let's go."

We were getting up high and the snow was getting deeper. Keith had to use every move he knew and I hung on grimly, steering into skid after skid, trying to anticipate his need to brake, then pumping furiously as soon as I saw his brake lights. Every switchback was an adventure; a couple of times I thought sure my rear wheels were headed over the edge. I could imagine tumbling off the side of one of these drop-offs and dragging Keith with me, both of us going end over end to the bottom, the towrope some sort of umbilicus tethering us on the highway to hell. Like something out of a Stallone movie, only they'd use a stunt man, I thought. And it would be happening to the bad guys, not Sly's character.

There was no time to be relieved each time we barely averted disaster; another crisis was always just ahead. I was sure we'd lost it around one corner. I turned the wheel desperately and cursed. Why the hell was he going so fast? When I got it straightened out I looked up and understood. The steepest part of the climb lay just ahead and Keith knew he needed every rpm he could get to pull me over.

Suddenly the snow flew out in a huge arc from his wheels; he'd just hit the biggest drift of the day. He slewed sideways and his speed slowed dangerously. If he comes backward, I thought grimly, he'll land on top of me.

Keith had barely enough momentum to get to the crest, and his rear wheels grabbed then and carried me on over the top. I gave him the thumbs up out the window and I saw him grin wide in his rear-view. It had been a very close thing.

He stopped when we got to the hill right above the road to the calving shed. "Nice piece of driving, Mr. Deal," I said as I got out. Keith just grinned. When Tyson pulled up behind us in the retriever, he rolled down the window and said, "That was quite a sight, you guys going over that hill."

We unhooked the tow rope. "Tyson, give David a push here and we'll see how far he can coast," Keith said. Not far enough. I came to a stop by Keith's house, and Tyson pushed me on up the hill and into the yard. I rolled to a brakeless stop up toward the barn, got out and went into the shop, shaking my head. "Quite a sleigh ride," I said.

"Yep," Keith said, already on to the next item of business. "We need an oil change and service on that crew cab. David, take the air hose and blow out the interior. Then vacuum it and clean the glass inside and out, and help Tyson with the rest of the service if he needs it." I got the message. No time to stand around and jaw about what just happened. Lots more needed to get done before it was time for that.

MARCH 30

Bringing in the drop—moving the heifers to the shelter of the alley behind the barn for the night—is one of my favorite jobs. It's close to the end of the day, and it's a task that proceeds at its own pace, which is to say the pace of irritated, very pregnant cows. Today, even though it was cold, it felt good to be outside after an afternoon working in the shop. The sky was mostly overcast, but the mountains to the west were rimmed with fire where the sun had gone down a quarter of an hour before.

Christian headed over to push the bulk of the heifers away from the manger, while Fletcher and I crossed the creek to get half a dozen stragglers lying in the far corner of the pasture. Only after we got close enough to touch them did they get to their feet, grunting and grumbling, backs and bellies impossibly wide, and grudgingly begin their nightly trek.

Back in the pulling room, I started to prepare a bottle of milk for the llama and got an end-of-the-day chuckle. On the side of the cabinet next to the sink, where the milk bottles were stored, was a sign printed in angry capital letters:

> WHOEVER'S CUTTING THE NIPPLES ON THESE BOTTLES STOP OR I AM GOING TO KICK YOU IN THE NUTS.
>
> KEITH

chapter 5 • . . . the crew . . .

"Ma, do cowboys eat grass?"
"No, dear, they're part human."
—CHARLIE RUSSELL

A COWBOY IS A STEREOTYPE THAT DEFIES ITSELF. The difference between romance and reality is too great. Today's cowboy doesn't fit neatly into any category, as I was learning—Willie John, Tyson and Christian, say, could not be more different from one another—but for that matter, neither did yesterday's. From the days of the open range, most ranch hands were not born to the work, but instead came to it from every conceivable walk of life. Historian Michael S. Kennedy cited a survey of fifty-five Wyoming cowpunchers in the 1880s that showed they came from nineteen states and three foreign countries. They had previously worked in freight hauling, law enforcement, mining, scouting, riverboat piloting, railroading, politics, soldiering, banking, smithing, plumbing, teaching and hotel keeping.

"Cowboy" is ubiquitous in our culture, a word that means much more, and much less, than its literal translation. Cowboys and Indians. Cowboy hat and cowboy boots (neither of which are used all that much on the ranch). Cowboys Are My Weakness. The Dallas Cowboys, who, being millionaires, are strange cowboys indeed.

Mama, Don't Let Your Babies Grow Up to Be Cowboys. A Cowboy's Work Is Never Done. (Amen!)

From Tom Mix to John Wayne to Clint Eastwood, cowboys have been deified, made into the most enduring heroes of a culture that worships heroism. In fact, cowboys are heroes, but not of the Hollywood variety. Their heroism comes in small portions. John Wayne may have saved the stampeding herd in *Red River*, but in real life the herd is saved one calf at a time.

Mostly it is hard work. Survival, not heroism, is the goal. The ranch crew functions much like some sort of native tribe, used to rigor, privation and isolation far beyond the experience of most people. That was true a hundred years ago, and it is true today.

There is that famous cowboy pride, pride in toughness, endurance, experience and know-how; pride simply in being a good hand, or "handy." It's no accident that the crew works twice the hours at half the pay that any unionized worker in any of their areas of expertise—a mechanic, a heavy equipment operator, a truck driver, a plumber, a welder, a construction worker—would put in. They do so because they are cowboys, and that's what cowboys do, even when they're not being cowboys, i.e. riding fences or roping wayward steers. Crew members depend on each other, and the chemistry of the crew depends on how often mutual trust is upheld and how often it is violated. No bullshit goes undetected for long; no hard work is unnoticed.

As I began to function as part of this crew, my respect for each member grew apace. Willie John was standoffish at first; Christian and Fletch and Tyson were enormously patient and helpful. Shared experience brought me closer to all of them.

One March afternoon I took a tour through the heifers with Fletch, and we brought in a bulbous, agitated black-and-white lady with a neat set of hooves protruding from her southern exposure. She required immediate attention. Willie John got a rope on her and Fletch and I did the pulling, and the little heifer calf popped out, easy as you please.

The experience of being involved with the beginning of a life is a hell of a rush, even if your back feels like somebody dropped a piano on it and you have cowshit and blood up to your shoulders. Fletch and I high-fived like a couple of kids, eliciting a contemptuous chuckle from Willie John, who growled, "Celebrate later. Get that calf tagged and shot." Which we did, not much chastened.

A little later I made my way across the creek and found a big red Angus heifer, No. 242, walking in circles, kinking her tail and kicking at her belly with a hind leg. That earned her a billet in the freshly cleaned Jug 18, where she circled agitatedly for half an hour before Keith said, "Willie and David, you better go give that red cow a hand." "Keep her looking at you," Willie said, and I did, long enough for him to slip up behind her and get a rope around her head. He flipped the end up to me and I took a wrap around a corner post and jumped down into the jug. "Easy. Don't startle her." Slowly Willie and I gained leverage until her head was snubbed up pretty tight against the corner, and I tied the rope off. Willie did a nice two-step to avoid getting kicked and reached up to see if he could get hold of the calf's back legs. He got chains around them, and he and I each pulled, sitting in the straw and bracing our legs against her. After a few minutes we were rewarded with a black calf. Willie swung it to clear its lungs, and soon it was breathing fine. "Go ahead and tag it and give it a shot," Willie said. "Something felt funny in there." He reached back inside the cow. "Son of a bitch. I thought so. There's another one."

"No kidding," I said. "Can I pull it?"

"Be my guest," Willie said. "Let me get the chains on, though; it can be a little tricky."

Thanks to his brother's exit, the second black bull calf slipped out more easily. Willie had already got a tag ready—"242-2"—and by the time he had applied it, reloaded the syringe with vitamin A and given the second shot, everybody appeared to be fine. Mom, understandably, seemed to be feeling much better, and she was already cleaning both calves. Even Willie seemed pleased. The next morning, when Tyson and I returned from feeding to clean jugs, we found a red-and-white crossbred heifer, No. 285, in Jug 4 by herself. She seemed upset, mooing loudly and turning tight circles in the jug, her calf nowhere to be found.

"I think Keith was doctoring that calf for pneumonia," I said. We looked around the barn to see if her calf had gotten out of the jug, but no. Then I walked outside and found the calf, dead. Keith had apparently dragged it out.

Just then Keith radioed for Tyson. "Hey, Ty, skin that dead calf and graft one of those twins to the mother, would you? Probably the second one, because the mom seems to be taking care of the first one a little better."

"Why can't the mother take care of both of those twins?" I asked. I felt quite proprietary toward those twins, particularly to No. 242-2, the one I had pulled by myself. "Well, she can," Tyson said, "but remember, these heifers are first-time mothers and their milk production may not be great. Plus one calf is plenty for a mother to look after."

So while I cleaned jugs Tyson went about the rather grisly task of skinning the back hide from the dead calf. When he was done, I held little 242-2 while Tyson tied the hide to his back with baling twine, looping it around his legs and body. We added 285 to his tag and put him in with the red-and-white cow, who sniffed him suspiciously. "Cows identify their calves by scent," Tyson said. "Her mothering instinct is obviously really strong. She knows something's wrong, but this often works. We'll see how she takes to this little guy."

"How long will he have to wear it?"

"Oh, two or three days ought to do it, if she's going to take him at all." By the afternoon, both bereaved mom and adopted calf seemed content.

On weekends Tyson and I fed hay with his brother, Wylie, a fifteen-year-old high school sophomore who worked weekends and summers on the ranch. Wy and Ty. Wy was a fair bit smaller than Ty, though still growing—maybe five-seven or eight, built wiry, sporting baggy snowboarder garb, a very cool haircut quite a bit longer than his brother's Marine sidewalls, and the same ready smile. First Tyson, then Wylie had come here for what Tyson jocularly called "Uncle Bill therapy." When teenaged rebelliousness struck, their mother had lived enough of ranch life—and knew her brother Bill well enough—to know that they would respond best under his care. Knowing Tyson as I did already, I could certainly see it was working. He had deep respect for Bill, and while college had so far not been an unqualified success, he had graduated from high school in White Sulphur, excelled in the Marine Corps reserve, and put in nine years of what we were doing right now, which spoke pretty highly of him in my book. I had seen Bill with him enough to know that as demanding as he was with all the crew, Uncle Bill was probably tougher on Tyson than anyone. No family favoritism dispensed here. Wylie had the honor of being named after his Medal of Honor–winning great-uncle, William Wylie Galt. I knew this already because his second cousin, Wylie Gustafson, a zany, talented red-haired country singer, had been a friend of mine for several years. Bill's name was also William Wylie; the family was justifiably very proud of his heroic namesake.

They'd be proud of this Wylie, too, I thought. It didn't take long for me to see that he was as hard a worker as Tyson, which was saying quite a bit.

Christian and Fletcher certainly made life easier for me during my first few weeks at Birch Creek. They hadn't been there all that long themselves—about six months—and they were happy to have someone to help with the menial work that came their way as junior members of the crew. Both were Southerners, both recent college graduates—Christian in English, Fletcher in biology and animal science. Christian loved to talk about fiction, poetry and music while we cleaned jugs or fed hay, and he was almost never without a quick smile.

In return, I tried to help lighten their load. I volunteered to take over feeding the hounds in the mornings, which I enjoyed. Usually I also started the four-wheel-drive retriever, which was a small pain in the ass because you had to crawl up onto the frost-covered brush guard in the front and balance on it while you opened the hood, checked the oil and scraped the windshield.

Christian was leaving at the end of March, headed back to South Carolina to take a desk job with his father's firm. One afternoon just before he left, the whole crew spent the afternoon at the shop, removing Lucy's engine.

Keith directed the procedure like Michael DeBakey doing a heart transplant, monitoring each step sharp-eyed: "Take the hood off, Christian. You and David carry it outside. Put it by the reefer. Tyson, get under there and get the hoses and the water pump and the fan and all that stuff. Fletch, take that fuel pump off and the lines." So it went all afternoon. Nobody said much. Keith was testy and exacting, roasting the crew when mistakes were made. Finally, with Keith and me underneath, shoving to disengage the splined shaft from the transmission, and the rest of the crew pulling from above, it was done. Lucy looked pretty mournful, grille bent downward, one headlight gouged out, a great hole and a puddle of oil the only thing remaining where her hood used to be. We would have to wait now for a donor heart to be shipped from the engine rebuilders in Great Falls.

Bill came in just as we finished. "Got it out, eh? Good work." Those two words lifted the atmosphere of the entire shop. Even Keith lightened up several shades.

On Christian's last day, he and Fletch got emphatically stuck in the Chevy retriever on the way into the O'Conner. I'd already noticed that there was a certain hesitation on the part of the unfortunate party to use the word *stuck* when calling for help on the radio. The usual protocol went like this:

"Keith, are you by the radio?" Keith knew damn well that a call from one of us early in the morning meant we'd screwed up. Twenty seconds of silence. Finally, desperation would win out over embarrassment and the call would be repeated. "Keith, are you by the radio?"

"What do you want?" Disgust would fairly drip from the radio receiver.

"Keith, we're on the road going to the O'Conner and we're, uh, we're going to need a pull." This was the euphemism of choice.

"How'd you get *stuck*?" No hesitation on Keith's part to call a spade a spade.

"We got down in one of these ruts and it just wouldn't go."

"Don't drive in the ruts."

An ominous silence. Then Keith would take pity on the wretched. "I'll be there in a while. I'm busy here."

Often Bill would follow up on the radio with some pointed advice about driving carefully and not wasting time and bringing the whole operation to a standstill. Or maybe he'd say nothing just then, and you'd think you'd escaped with just the butt-chewing Keith gave you when he came to tow you out, but then at lunch Bill would suddenly say, "Who stuck that retriever this morning?" and you'd get both barrels after all.

Bill had an uncanny way of knowing everything that happened on the ranch. The radios were partly responsible, but only partly. Of course the flying helped, too. Bill had two airplanes, a Cessna 185 Skywagon and a little Piper Supercub, and he was forever flying over the ranch and spotting this or that that needed doing or was being done incorrectly.

Both Keith and Tyson had been surreptitiously chewing tobacco for years. Bill was a reformed smoker and chewer, and he would have given both of them endless grief if he had known of their habits. Several times already I had seen close calls with both of them, when they had barely managed to get rid of chews after Bill popped up unexpectedly.

In April the crank on the driver's-side window of the diesel retriever stripped out, with the window in the down position, naturally, and so we had to put cardboard in the window until parts arrived. And all during that period, when he was chewing tobacco during feeding, Tyson had to open the door to spit. He just knew that some morning Birch Creek One would be flying overhead and radio, "Why do you keep opening and closing that door?" But somehow he still escaped detection.

The window handle was the least of our repair problems. "You guys have got to treat these outfits better," Bill said at lunch on Christian's last day. "We've got a shop full of broken-down vehicles. Willie John, you're breaking four-wheelers too often. A new engine for Lucy's going to cost me twelve hundred bucks. David, was that you driving the diesel out of the horse pasture this morning?"

Startled, I nodded.

"You were revving the engine way too high. That's hard on engines and transmissions."

"Okay, Bill."

What Bill was really distressed about, we knew, was what had happened once Fletch and Christian had gotten unstuck. As they were feeding hay with the truck on bungee autopilot, they had run over a calf and killed it.

I knew Christian well enough to realize how much it was bothering him. It was easy to do, with the mothers crowding around the retriever and the calves trying to stay close to them, but Christian had spent the past few months trying so hard to bring calves into the world, and to save them once they'd arrived, that to kill one was a crushing blow. Bill had shaken his head resignedly when Christian had told him. He'd lost dozens of animals to the weather, and it didn't help to lose one like this.

When Bill gave Christian his final check, he tried to wave it away. But Bill said, "I don't think that's a very good idea. You earned it, take it." There was affection apparent in his gruffness. He knew how hard Christian had worked, despite his youth and inexperience. I could see that he enjoyed having the two college grads on the place, for their enthusiasm and work and for their conversation. Their months of working together had produced a bond that I could only hope I'd have the chance to achieve.

That same day I got my first paycheck. I'd worked fifteen days in the pay period and got a check for $389.30. Well, I thought, most people who do this don't have a mortgage. They've got the bunkhouse, free of charge, and that's enough. Strangely, that check meant as much to me as any paycheck ever had, even the ones with a couple more numbers on them. I never thought I'd earn a dollar this way, and I was pleased to find out I could.

I could tell that Keith, too, would miss Christian. He joked with him about getting drunk and laid when he finally got back to civilization, and after work, Keith spent a few hours of his own time putting a clutch in Christian's Bronco so he could realize those two goals. Fletch was saying adios to a close friend, roomie and workmate. He would be here another month or so, and after that he would go back to the dude ranch where he and Christian had worked last summer. The three of us cooked dinner together this last night, and Christian left me with paperback copies of a couple of favored novels—a rather disparate pair, *The Remains of the Day* by Kazuo Ishiguro and William T.Vollman's *Whores for Gloria*. I appreciated both greatly; whenever I could stay conscious long enough in the evening, a good book felt like a real treasure. So did a friend; Christian would definitely be missed.

With Christian gone, I became Fletch's feeding partner. The first week of April it turned cold again, and the northwest wind glazed sheets of ice over every exposed surface, including roads, hay bales, stackyard panels and a few more frozen calves. The rest of the cattle clumped together, trying to find a windbreak, and feeding was nightmarish.

One morning we were on the Gurwell, getting alfalfa for the yearlings. Fletch backed the retriever to the stack, set the teeth under the load and tried to raise the bed. The bales were frozen to the ground and something had to give. *Crack! Crack!* Like two nearly simultaneous rifle shots, the chains on the sides of the bed broke.

Fletch pulled forward and lowered the bed. We looked at the chains lying there, the same thought in both of our heads. Fletch voiced it: "Oh, shit, we're in for it," he said wearily. He reached for the mike. "Keith, are you by the radio?"

It was twenty minutes before he could get his service truck there and another forty-five before the three of us could get the chains repaired satisfactorily. "They have to be exactly the same length, and they'd better be on there good," Keith said. "If they break at the wrong time they could kill somebody."

He was good-natured about what happened; I think he could see that the weather had produced an unusual circumstance. As he expertly pounded the links into place, though, he warned, "If a load won't come, back off right away. You could break the frame or the bed. They won't take that much torque." He inspected the frame for cracks and fortunately found none. As he did so, he told us that Willie John had cracked a frame trying to lift a frozen load a couple of years ago. I think it made Fletch feel a little bit better to know that what happened could have been a lot worse.

As we drove back from the Gurwell, I noticed a noise that wasn't quite right. "What's that clanking noise?" I asked Fletch. "The truck," he said, deadpan, and for some reason that seemed like the funniest thing I'd heard in months. Thankfully, the noise turned out to be nothing worse than a broken link on one of the outside tire chains, which I took off, fixed and reinstalled when we got back to the shop.

I was not the junior member of the crew for long. A few mornings later, as Tyson and I got ready to go feed the bulls, a little Nissan came hurtling up the driveway. "Slow that thing down," Tyson muttered. I thought it was fortunate that Keith and Bill weren't present, or there would have been more than a mutter.

Out stepped a rangy, slim young man in jeans, cowboy boots and a red quilted jacket. "You must be Jerry," Tyson said. "Yep," said the kid, smooth-faced, handsome and smiling. He looked maybe nineteen; Jerry Huseby was in fact eighteen, and Bill had just given him a job.

"Hop in," Tyson said. "We've got to get going." It was cold and windy; it felt like snow was possible.

Jerry was barehanded. "Did you bring gloves?" I asked. "No, not today," he said. "I've got to get some when I can."

"Feeding hay is pretty tough without them," I said. "Take these. I've got another pair."

"Thanks!" He flashed me a relieved smile.

I learned a few things about Jerry that morning. He was born in Harlowton, due east of White Sulphur. His dad worked on ranches all the time Jerry was growing up. He loved to ride. His big brother,

Joel, worked at the sale yard in Billings, and rodeoed some. Jerry had most recently worked at Castle Mountain Ranch, another of the largest ranches in the area, along with his father and brother. He liked old Ford trucks. He had a girlfriend, Shyla Low, a White Sulphur Springs high school senior.

Jerry was pleasant and obviously happy to have found a job, though not particularly industrious, and for a Montana ranch kid he didn't seem to know very much. He also seemed dog tired all the time. He slept in the retriever between loads, and he fell asleep right after lunch in the corner chair of the living quarters, with Cow Cat, an elderly tabby who frequented the barn, on his lap. Willie looked over and chuckled, then filled a large syringe with water and squirted it at Jerry. That got rid of Cow Cat, but it took a second syringe to wake Jerry.

Then when I started to do the lunch dishes, Keith said, "No, let the new guy do 'em."

Jerry laughed uneasily. "I, uh, I don't know how. My mom always did the dishes."

Willie John routinely pulled rank and refused to do the dishes. I didn't have much problem with that; Bill and Keith didn't do them, either, and Willie was a senior hand. But when Jerry said he wasn't able to wash dishes, I just gaped at him, and so did everybody else. Keith recovered first. "Well, Jerry, it isn't hard," he said with some bite. "Fill the sink with hot soapy water, put the dishes in there and wash them. Rinse them off when they're clean." He rolled his eyes at me as he walked out of the living quarters.

At that rate it didn't take long for Jerry to dig himself a real hole with Keith and the rest of the crew. He was frequently late for work. "Overslept" or "Alarm didn't go off," he'd say with a grin, and then stoically accept the ass-chewing that would follow. Working with him was aggravating. He was always the last one out of the truck, and he never hurried to get a gate, to fill the manger, to clean the floor. He always had to be told to get moving; Keith, red-faced, yelled at him for ten minutes one morning when he caught him leaning on his rake talking about rodeo instead of helping us clean jugs. Willie was particularly disgusted, perhaps because Jerry was a native Montanan, a young, strong hand who'd grown up on a ranch and should have been useful.

I began to understand, by example and the lack of it, the real cowboy ethos: Don't duck out on your share of work. Don't make somebody else clean up your mistakes. Take care of the animals and your fellow hands.

Even as the cowboy on the ranch is thought to be automatically imbued with all sorts of Boy Scout virtues, like bravery, gallantry and jut-jawed determination, the prevailing stereotype gives him a little latitude when he finally gets to town. Then he is expected to be rowdy, to drink red-eye whiskey, to fight and fuck and play stud poker.

Well, yes. There is reason for all of this. Historian Lewis Atherton wrote about the days of the open range: "The cowboy's life involved so much drudgery and loneliness and so little in the way of satisfaction that he drank and caroused to excess on his infrequent visits to the shoddy little cowtowns that dotted the West. A drifter whose economic status made it difficult for him to marry and rear a family, he sought female companionship among prostitutes." Things are a little better for today's cowboy, but drift and drink are still common, and by that standard I reckoned Bill Galt was pretty lucky. He agreed. "When I was growing up on ranches, when I worked on them as a kid, drunkenness was the rule, not the exception. All the hands drank."

The current Birch Creek crew's forays into White Sulphur were pretty tame, and also infrequent, considering how close we were. Tyson, Fletch and Christian and I had eaten dinner a couple of times in town and had had a drink or two, but usually nobody wanted to stay out too late. And if I had a day off, I would usually drive home to Livingston.

So one evening when we finished work at a reasonable hour and Willie suggested a few games of pool, I accepted readily. Almost every town in Montana has a bar called the Stockman or the Mint. White Sulphur (and Livingston) have both. We drank whiskey and played pool and ate an astounding number of the Mint's excellent cheeseburgers. I was just as handy as Willie John at all three tasks, and that was a rare and enjoyable feeling.

Different ages, backgrounds and temperaments notwithstanding, it seemed like we were becoming friends in spite of ourselves. That was the biggest surprise of this new life: I'd never made so many good friends in such a short time. Being a tribesman will do that to you.

As we headed back to Birch Creek with a morning hangover guaranteed, Willie John fretted a little about getting back late. His wife, Erin, was pregnant, and he knew she'd be irritated. He was more talkative than usual, and he told me about working at Lingshire for Bill in past years, then moving to Billings and working for the state as a brand inspector. "Great hours, weekends off, but I missed being on a ranch," he said, and that was the paradox I knew a lot of hands faced. A town job was usually more money and less work, but if you liked this life, what were you going to do?

APRIL 3

Tyson, Wylie and I took the diesel retriever up to the O'Conner to feed this morning. Tyson flipped on the radio, which was tuned to a Great Falls station. Rod Stewart had barely begun croaking about Maggie May before Tyson started punching buttons.

"Ugh," Wylie said. "I get sick of Rod Stewart. Bill listens to him all the time." Tyson nodded in vigorous agreement. But the Helena station he turned to was right in the middle of "Do Ya Think I'm Sexy?" Which was bad enough, but the last station available, also out of Helena, was no help. "Tonight's the night," crooned Rod.

Tyson stared at me slack-jawed. "Whoa. That's never happened to me before."

"Me, neither," I said. "The Rod Stewart Bermuda Triangle. I wonder what it means."

"Man," said Wylie. "We could have a really weird day."

chapter 6 • . . . and the greenhorn

"When you call me that, smile."
—OWEN WISTER, *THE VIRGINIAN*

IN APRIL I DREAMED OF CLEANING JUGS. I dreamed that I walked into the calving barn after feeding and the dirty jugs stretched from the door eight miles due east, all the way to White Sulphur Springs. I started cleaning them, piling the dirty straw higher and higher on the back of Lucy until it was close to the roof of the barn. Keith would come up over the horizon on a four-wheeler to check on me every once in a while. "Hercules didn't have this much to clean at the Augean stables," I told him with an earnestness only possible in dreams, and he nodded wisely in agreement, then said, "Hurry up now, you've got to get these done before lunch."

It seemed just like that, most of those fading winter days, shitwork to the horizon. I wake at six A.M. to hear the bulls demanding breakfast in the corner of their pasture just across the road. One of them—I know from feeding that he is a smallish red Angus—has a curious high-pitched bray, like a countertenor. I clear my head with coffee, clear the windshields with a scraper, dump another quart of power steering fluid in the diesel, feed the hounds and go feed frozen pigs of bales to the loudmouthed bulls and the equally demanding yearlings. Clean the goddamned jugs. Sweep the shop, change the oil on something, bust tires, pull a calf, fill the manger. Hose down the pulling-room floor for the fifth time today and probably the two hundredth time this month. Clean the mud out of the drain. Mop the living quarters. Cook lunch and do the dishes. Mop again. Check the drop again. Clean the floor again. Fill the manger again.

Shitwork.

Occasionally Keith will come up with some new mind-numbing torture. "Go clean out that oat bin, the one second from the left. I want it *perfect*. Sack up the good oats and clean up any moldy oats. It's got to be spotless. They're delivering oats tonight and all that mold has to be out of there."

Clamber inside the huge bin. Shovel and sack the good grain, not too bad. Near the door some moisture has intruded, and the oats have stuck to the concrete floor and metal sides of the bin, over an area probably ten feet square. The black moldy grain is rock-hard; it resists repeated hacking with the shovel blade and the shop broom. Finally, on my hands and knees, I find a wire brush somewhat effective, but the brush is so small I feel like I'm polishing a parking lot with a toothbrush. In a few minutes, of course, Keith's head pops into the hatch opening, rasping voice reverberating around the stark empty metal and concrete: "Goddammit, hurry up in there! That shouldn't take you all day!"

From oat bin to ash can. Burn barrels are scattered in twos and threes around the place: by the chute shed, in front of the calving barn, near the shop, near both trailers, at Keith's place and at Bill Galt's house. I know this because after I manage to finish chipping moldy oats, I burn trash, then take the stinking barrels of ash and blackened cans to the dump.

I hoist the barrels into the back of the old maroon pickup, half a dozen per load, then drive them up above the shop to the other dump, which is up above the shop, on the edge of a field called the Dry Gulch. The dump I went to with straw each day is basically organic—the gutpile, it's called, with its dead animals and heaps of straw. This one is more like a typical landfill, a large pit half filled with garbage burned and unburned, old tires, refuse from the shop, a few items of ancient ruined furniture and such.

Always, there were more jugs. Christ, these heifers managed to shit on a lot of straw. Forking out old straw. Raking. Scraping. Forking in new straw. Forking in hay. Hauling water. Running cows and calves up and down the alleyways. Kicking pairs out into the pens behind. Checking the drop, bringing more cows in. Shitwork.

"David, when you get done with that last jug, take that big-titted cow out of No. 8 and put her in the head catch and milk her, then give the milk to her calf."

"Okay. Why isn't the calf able to suck?"

"Her tits are too big," Keith said. "He can't get 'em into his mouth. He's getting a little from her but not much."

So I started this Mamie Van Doren of cows down the alley toward the door into the pulling room. "Get right behind her," Keith yelled. "You've got to push 'em hard once you start to the chute or you'll have a wreck. Tyson, go around and get the door."

I ran her the rest of the way in and got her a bucket of grain to keep her front end occupied while I was busy with the other. Tyson detected a little hesitation on my part, and smiled. "You ever milked a cow? Didn't think so. Get a bucket over there, and one of those wide-mouthed bottles. Turn the bucket over and sit on it. Okay. Set the bottle there. Now squeeze the top of the tit, then close the rest of your fingers down on it one at a time and pull, like this." A prodigious stream of milk hit the bottom of the bottle. I did the same thing, and didn't get a drop. Tyson laughed. "It takes some practice. Keep trying. You'll get it. Be gentle, though. She can sure kick you into next week when you're in that position."

Before long I was getting a respectable return on my efforts. Tyson grabbed another bottle and another tit, and pretty soon we had them both full.

"See how chapped that tit you've got is?" Tyson said. "Must be from him trying to suck, I guess. Get some of that Bag Balm up there above the sink and put it on. Or the other jar there, the Teat Elite. I like that stuff better, actually." As I did so, Tyson said, "You know, that stuff's great for your hands, too, especially when it's dry and cold like this. Use it every day and it'll help." I was grateful for the tip; during the very cold weather my fingers had gotten chapped and cracked.

Perhaps as a statement about my technique, the cow disdainfully shat all over us and, worse, all over the freshly cleaned and disinfected floor. "Shit!" we both said in unison, and then snorted with laughter at the accuracy of our epithet.

We left Mom in the head catch for a few minutes while I fed her calf. I found the easiest way to do it was to straddle the little guy and lift his head up. Once he got the idea he was enthusiastic. As I led the cow back to the jug she mooed nervously, swinging her head around to find her calf, and he answered her. When she spotted him she went over and sniffed at him, no doubt picking up my scent, and wheeled around irritatedly toward us, her considerable bulk planted squarely in front of the calf. As I went to finish the rest of what needed to be done—clean the floor again, feed the baby llama—I wondered how this madonna and child would fare in the next year.

Keith had marked her on the hip with big letters in green paint: BT. It took me a few moments to realize that stood for well-endowed. It was not a frivolous mark; a cow that couldn't be sucked wasn't a whole lot of use. She was destined for an early trip to the sale yard. Too bad, I thought. She was a good-looking young red Angus cow. As for her calf, he was going to need a lot of help for the next few days. And once they were out in the pasture, away from people, he was going to have to be able to turn that faucet on himself, or else. Right now I felt damned near as helpless as he was. After fifty days of the same routine I was beginning to wonder what the hell I was doing, and how many more days it would take for me to become self-sufficient and productive around here.

I could not remember being immersed so totally in alien circumstances. It felt like what I had always imagined time travel would be like. How often do you parachute into a completely different life, not doing the same thing in another place or for another company, but working and living in an utterly new way?

This world was exotic in some ways and stripped-down lean in others. After you have worked twelve hours, your evening becomes an agenda of survival without frills. Wash work clothes and hang them up quick or they will not dry—particularly the quilted coveralls that make outdoor existence possible. Cook something to eat and eat it. Wash the dishes. Take a shower. Write a letter, make a phone call. Sleep, because you must be up an hour before dawn to do it again.

Life is thus reduced to essentials. Work, food, clothing, cleanliness, shelter, maybe a quick burst of contact with the world. No strolls downtown, no cocktails with friends at the bar—my night with Willie John at the Mint and dinner with the rest of the crew being the only exceptions to date—no newspaper, no CNN. So, isolation intensified the numbing sameness of the work and the dreariness of the season—by its sixth month, winter has worn out its welcome—and for a while I managed to feel pretty sorry for myself.

Somewhere in that dismal stretch of days, I managed a personal milestone: I picked up my first load of hay with the retriever. Tyson had to go load hay that morning for Doc Schendal. Doc bought hay from Bill and came over every few days with a little trailer, which Tyson or Keith or Willie would load up using the forks on the front of the big yellow Dresser loader, or if it was in use, the little Bobcat skid-steer. So when Keith radioed and told Tyson to go load Doc one morning, he told me to pick up a load and drive it down to the calving shed to wait for Ty.

I'd probably seen it done a hundred times, but actually doing it was something else. I went through the steps carefully: Back square to the stack. That took a few tries. Engage the PTO. I had to bend over and pull the red knob with both hands, and even so it took awhile. I was red-faced by the time I got it, just as Tyson usually was. Open the rear forks. Raise the bed most of the way, until those forks were parallel to the ground. Back up to the stack another couple of feet. Raise the bed a little more, then tuck the rear forks in halfway. Raise the bed the rest of the way and back up so that the entire bed was flush against the stack. Put the front tines down. Put the rear forks in all the way. Start the bed back down, just a few feet. Stop to make sure the load was coming—it was. Pull up a little to give the back end clearance to pass the rest of the stack. Lower the bed the rest of the way. Disengage the PTO: more bending over and huffing to get the balky cable back in. Then pull away slowly. Success!

It seems like a little thing, looking back on it, a relatively simple procedure that was done perhaps a dozen times a day. I would do it hundreds of times more. But it was a big thing to me that morning. For a few moments, at least, I felt useful.

The shop was probably the most humbling venue on the ranch. At least working outside, even in the weather, gave me some visual pleasure. But many of those frozen days, the shop was where we ended up.

"Tyson, David, you guys get on that fancy truck," Keith said one afternoon, so we turned our attention to Bill's new pickup, an elegant white three-quarter ton straight from the Ford dealership. This was his going-to-town pickup, not the one he drove around the ranch every day, but it still needed to be rigged out for work, with a two-way radio and a more rugged set of tires than the puny ones mounted at the factory.

The tire shed gave a pretty good hint of the size of Bill Galt's operation—or at least how mechanized it was. Piled floor to ceiling were truck tires, tractor tires, semitrailer tires, tires for the retrievers, for the four-wheelers, and for machinery I didn't recognize. Tyson and I rooted through the stacks until we found a set of sixteen-inch Toyo all-terrains for Bill's outfit. They were the tire of choice around the ranch, for good reason. "They'll turn the sharp shale we get out here, most of the time," Bill said. "The rocks go right through the rest of them." He watched me struggle with the tire-mounting machine and snickered. "Quite the mechanic, aren't you?" After everyone had a good laugh, I further impressed Bill by trying to mount one of his wheels wrong way out, which caused a new round of guffaws.

A few days later, I spent a morning helping Tyson and Willie rehabilitate the old maroon pickup. While they worked on the brakes, I was assigned the task of installing new shocks on the front. When

I got the top of the first shock bolted to the frame, I said to Tyson, who was by now installing a new fuel pump, "Jeez, they ordered the wrong shock. This one's too short."

Tyson took one glance, then looked at me with a grin that was both pitying and gleeful. He went to find Willie John and pass on my latest mechanical gaffe. Turned out I had failed to cut the retaining wire that kept the shock compressed in its box. Red-faced, I snipped the wire, added the washers and bushings for the bottom and bolted it in, then did the other one. As I drained the crankcase and went to the reefer to get oil and air filters, Tyson said, "Did you get that shock stretched out, Mr. Goodwrench?"

I finished the oil change and spent an hour cleaning the yellow Kenworth cab and sleeper, wondering if I would ever get to go anywhere to work in this truck, or if I would always simply clean it for—and after—someone else. I took the air hose and a pressure gauge outside to check all eighteen wheels, and as I crawled around the trailer—the inside duals were a bitch to get at—I felt like Jimmy Stewart, the lawyer washing dishes in *The Man Who Shot Liberty Valance,* with John Wayne laughing at him and calling him Pilgrim. Pilgrim. Greenhorn. Tenderfoot. Honyocker. The rather xenophobic West has always had plenty of words for those who would presume to crash the party, to move in and try their hand at being Western. I am a lifelong Westerner, but suddenly I felt like an interloper. Yesterday's greenhorn is tomorrow's old-timer, I reminded myself. After all, Jimmy Stewart did manage to become a senator, with a little back-alley help from the Duke, and by God I could learn to be a good hand. And I liked "Mr. Goodwrench" a lot more than "Pilgrim."

APRIL 10

As Tyson drove into the steer pasture, going very slowly because the road was very slick, I noticed something odd out my window. "Hey, we're being followed."

Tyson looked at me quizzically.

"By what appears to be a pissed-off pheasant. I'm not making this up."

"Oh, wow, it must be Super Chicken." Tyson stopped the truck.

"Who?"

"Super Chicken. Bill tried to get a pheasant population to take hold here, but the coyotes got most of them. That little guy survived, though, and he's never been scared of people. He'll come and check you out every once in a while. Willie John caught him once and gave him a ride on a four-wheeler. Now, that really pissed him off."

Sure enough, when we got out he waddled right up to us, handsome tail feathers dragging in the snow. He wouldn't quite allow himself to be picked up, but each time I put my hand down he'd come up to me, then try to peck at it. He followed us all the way to the gate into the heifer pasture, then stomped off, toward the creek.

Part Two

new grass

Bald eagles back in the cottonwood tree

The old brown hills are just about bare

Springtime sighing all along the creek

Magpies ganging up everywhere

Sun shines warm on the eastern slope

March came in like a lamb for a change

Gary's pulling calves at the old stampede

We made it through another on the northern range

—IAN TYSON, "SPRINGTIME"

chapter 7 • gods with wide pure hoofs

AS I RETURNED ONE AFTERNOON FROM AN UNEVENTFUL TOUR of the grumpy mothers-to-be, the front doors to the barn opened and a muscular, compact man in coveralls walked in, leading two sorrel horses. They strained at their halters and their eyes looked wild and they snorted steam and switched their tails as they took in their new surroundings.

"Cut that shit out," the cowboy said, and firmed up his grip on the lead ropes. The horses settled some, gleaming red in the gentle barn glow, and as he led them into the last jug in the back of the barn and tied them to posts I caught myself marveling at the power of their bunched muscles, energy so close beneath the surface, and at the evident reluctance of their relinquishment of control to this man. There was something timeless about it, something remembered, something the same as other horses and other men, from Arab to mustang, Mongol to Sioux. After all, horses outrank us by seniority—eohippus was here millions of years before the Ice Age—and they have never taken particularly willingly to subjugation by man. I remembered some lines from a poem by Pablo Neruda:

Suddenly, led by a man,
ten horses stepped out into the mist.
Hardly had they surged forth, like a flame,
than to my eyes they filled the whole world,
empty till then. Perfect, ablaze,
they were like ten gods with wide pure hoofs,
with manes like dreams of salt.
Their rumps were worlds and oranges.
Their color was honey, amber, fire.
Their necks were towers
cut from the stone of pride,
and behind their transparent eyes
energy raged, like a prisoner.

"I'm Donnie Pettit," the cowboy said, and I returned his quick smile as I flaked some hay for mama llama. I walked over and he stuck out his hand, and I noticed that the smile extended to his pale blue eyes. A few other things stuck out: an early crop of gray hair from his red wool cap, his lower lip from the wad of snoose deposited there, and a round can from the rounded butt of his Wranglers.

"Hey, Round Ass, how you doing?" Bill Galt boomed. "Fallen off any lately?" To me he said, "There's an anvil and a big wood block and a box of shoeing nails over by the door to the pulling room. Bring 'em over here, please."

As I lugged those items across the barn, he said, "You know why I call Pettit Round Ass, don't you? I mean, besides the fact he has one? A round ass is someone who gets bucked off a lot."

Pettit snorted. "I know who's broken more bones falling off horses."

"Well, you've got me there," Bill Galt said. "A dozen that I know of." Willie John and I pulled a calf while Bill and Donnie were shoeing the horses.

"Come on, Pettit, I've got three legs done and you're not even done with the back," Bill said. "David, dump every goddamned one of those shoeing nails out and sort 'em. The sixes are in with the fives and the eights. And don't you leave a nail on the ground. A horse or a cow steps on one, it could ruin them."

As I stared at the mound of nails, Donnie said, "She's a little more of a pain in the ass than Roscoe." The filly he was shoeing was just three years old and not at all sure that she wanted footwear. "Come on, Susie, hold still," Donnie said, shoeing nail between his teeth. "Ouch! Goddammit!" He raised up and swatted her hard across the rump. She had jumped at a critical moment, and he had run a nail into the ball of his thumb. He shook his head, bent down and grasped her hoof between his legs again. She made precisely the same move, and he put another hole in his thumb about a quarter inch from the first one.

Donnie sighed. There was no other way to do it but to let her calm down a minute, then bend and pick up the same hoof again, knowing that in a few moments he could well have the same nail in the same thumb all over again.

Bill Galt finished with Roscoe, and as I sorted through my shiny mound of nails—the fives were quite hard to tell from the sixes, but I was getting good at it, albeit feeling more like a hardware clerk than a ranch hand—I said, "You don't like to contract much work out off the ranch, do you? No vet for the C-section, no farrier for the horses?"

"That's right."

"Do most ranchers do it that way?"

"Not so much as they used to," he said. "That's one reason why they have trouble. Some of these fellows aren't even ranchers any-more, really. They've quit doing the stuff that makes them money."

"Like hiring me to sort shoeing nails?"

"Hey, you're not killing any animals or breaking anything while you're doing that. And you're cheap. Keep sorting."

"How did you learn to shoe horses and do C-sections?"

"Of course I've been around horses all my life, but a lot of it I learned when I took a farrier course at Montana State. And I went to Fort Collins and did a seminar on C-sections. Then I came home and killed everything I touched." He smiled grimly. "I had to figure out how to do them right instead of the way they taught me."

The horseshoeing was a very good sign. It meant new, more interesting work ahead—specifically, trailing the older cows from pasture here at Birch Creek to their summer range at Lingshire, where Donnie Pettit and Pat Bergan would oversee their calving. These girls had been through training for this; when they were heifers, or two-year-olds, they calved in the barn, with plenty of help. When they were three, they calved up on the O'Conner, out in the pasture but pretty close by, and they got checked carefully each day. When they were four and older, they joined this herd and were trailed up to the big Lingshire spread, where they'd be on their own quite a bit more. Donnie and Pat would check them periodically, but by this time they pretty much knew what to do, and the cycle was repeated each year.

The Lingshire country, in many ways the jewel in Wellington Rankin's crown and now in Louise's, was perfect summer range but not so great for wintering animals; the terrain and the weather were too rough. So it made sense to move the cows from the relative shelter of Birch Creek, where they could be fed easily, out onto the grass at this time of year. I just hoped I'd get a chance to help push them up there.

The next day, Bill, Donnie, Pat, Keith and Willie started the cows on their way. It was more than thirty miles, and the drive would take three days. We caught another snowstorm on the second day, so the cows stayed parked at a little pasture Bill owned, called Candy Dan's, after homesteader Dmitri Candea.

The day the weather cleared, Bill called on the radio during feeding. "Tyson, is David with you?"

"Yep."

"Tyson, you finish up with Jerry. David, take the other retriever back to the shop and park it. Willie John, pick him up there. Hook up the white trailer and load Bose and John and Dusty. Bring an extra saddle—you'll find two in Donnie's truck when you get there—and head to Candy Dan's. Saddle all of them, and if I'm not there, leave Dusty for me."

"Okay." It sounded like I might actually get to ride for my pay today. I zipped into the bunkhouse and changed into cowboy boots and helped Willie hook up the trailer and load the horses. We were halfway to White Sulphur when I said, "Willie, did you load that saddle before you picked me up?"

"Fuck! Hang on, this is going to have to be quick. He'll be calling any minute to find out where we are." He flipped around—not so easy when hauling a big trailer, but he made it seem so—and high-balled it back to the ranch. The tack, I discovered, was kept in an old shed beside the chute shed. "Grab that one," he pointed. "I'll get bridle and blankets. Hurry up!"

Willie roared down the icy gravel like a moonshine runner with revenuers on his tail. "Anybody coming on your side?" he asked as we approached the highway. "Clear right," I replied, and he sailed into a quick, looping left turn. In a few minutes the radio made me jump. "Willie John, where are you?"

"Just passing the river field, Bill."

"Okay, hurry along."

"Okay." We *were* hurrying along at the time, to the tune of about 90. Willie John actually slowed to a sedate 75, now that Bill had fixed our whereabouts.

About twenty minutes later we pulled up to a set of corrals right by the road. Bill was standing by his truck, and Donnie Pettit and Bill's other Lingshire hand, Pat Bergan, were already ahorseback. Pat had opened a gate as we drove up, and he and Donnie were pushing a bellowing mass of cows through to the road. Willie opened up the trailer and led the horses out, handing one lead rope to me. "Tie him up to the side of the trailer and saddle him," he said. This was John, a stout sorrel fellow of a dozen or so years who was good-natured enough for someone with my marginal riding talent. I knew this because Fletch has ridden him before and had reported to me that he was rocking-chair easy but liked to go when asked.

Willie John made short work of saddling the other two. Bill mounted Dusty, a big black gelding, in a quick fluid motion. "You two take the drag," Bill said. "Push them along. Can you get on?" This last to me, eliciting a laugh from Willie John. Pat and Donnie seemed pretty damned jovial too, as I put my left foot in the stirrup and swung myself aboard, none too gracefully. "Jesus, just like a farmer," Bill said. "Don't hurt yourself." He wheeled Dusty around and rode to the head of the bunch with Donnie. Pat put his horse in the trailer; he would follow us so the trailer would be there for us at the other end of the ride.

Predictably, the "drag" turned out to be the back of the herd. I was eager to do something; as soon as the last cow got onto the road, I dismounted, closed the gate, got back aboard John and started pushing the cows. "Hey, David, wait a minute," Willie John called to me. I turned around and trotted back to where he waited with his horse. "Let them get strung out and going first. They're all balled up, looking for their calves. If we give 'em a little room they'll mother up and get lined out." We did, and they did.

The sun was bright and there was no wind. Particularly after the weather we'd just been through, it looked like spring. I knew better, but it was a stunning afternoon, like the first moment of falling in love, and it made you want to believe it would last forever.

Pat Bergan handed me up a sack lunch from the window of the truck, and I ate a sandwich, an apple and a candy bar with one hand and steered John with the other. There wasn't that much steering to be done along here; occasionally a cow would stop in the ditch and turn and gaze back down the road, and at us. Depending on which side she was on, Willie or I would ride down behind her and get her going again. Here at the drag were some cows with new calves and the cows who were slower for some other reason. A few were clearly sore-footed; some had some age on them, bony through the hips even in pregnancy, and others looked like they were just about ready to calve.

This unforgettable afternoon was worth frozen mornings and wrenched muscles: The feel of John, strong and willing, picking his steps with care on the icy road. The mellow, pleasant smell of cows and clean air. And sweetest of all, a rhythmic, repetitive symphony of sound: clacking of hooves, bawling of calves, the low-pitched moos of mothers reassuring them, the soft purr of the truck and trailer behind us, and our own exhortations to the cows: "Hike out there, come on, Mama, step up, step up, move along. Hoi! Hoi! Move it out, let's go, head up there, come on, let's go." I was actually a little too warm with the sun on my back—now that was a first—and I unbuttoned my denim jacket. I felt so comfortable I nearly dozed off a couple of times, and I suspect that if I had, John would have carried on just fine.

I pulled back and walked John next to the truck for a while so I could talk to Pat. I'd heard much about him. He'd worked on this ranch for a good thirty years, and he was respected as the cowboy's cowboy. He certainly looked the part; weatherbeaten handsome, with Paul Newman eyes and maybe three days of white stubble. He spoke with the calm courtesy of a man who has nothing to prove, his deep slow drawl a perfect canvas for dry wit.

"Incredible country, isn't it?" I couldn't help gushing as we pushed into the vast rippling foothills of the Little Belts, a little sagebrush poking through the snow on the nearer slopes, folds of unbroken white against the horizon.

"Yup," Pat Bergan allowed, and then waxed expansive: "Big country." A couple of times I had to get John into a quick trot to get around a straying cow, and Pat watched me with faint amusement as I hi-yied after these stragglers. "Sometimes it's best to just let 'em go. They'll be okay," he counseled. It was true; most of these girls were on autopilot. This trip was an annual ritual, and they knew where they were going, which was more than I did. A few calves who were still separated from their mothers kept turning back. "They're trying to go back to the place where they sucked last," Pat said. "That's their instinct." At least they gave John and me some honest work.

Here, at last, was a chance to connect with the roots of the cowboy, the raison d'être. This was the way cattle had been moved since Charles Goodnight and the other big Texas ranchers first brought their herds up to Montana more than a hundred years ago. Cowboys were stationed at the point, or lead; at the "swing" alongside the herd; and at the back, or drag, where we were. Admittedly, those boys didn't have Pat Bergan backing their play with a fancy new pickup, but otherwise the process was just the same.

So it went for three hours or so. The wind began to kick up as we moved northward, and the sky began to turn quickly from blue to gray as it only can in Montana. Within a mile or two that surplus warmth was replaced by a stiff, cold crosswind out of the west, driving little frozen spicules of snow into the left side of my face. John shook his head irritably and I quickly buttoned my jacket. In a few more minutes we reached a cattle guard and a gate posted with a sign: No Vehicles Past This Point/Galt Ranches. I figured they didn't mean John. As we pushed the last of the stragglers through the gate, one of the cows lay down just inside the fence and calved on the spot. I watched her for a few minutes to make sure all was going well, and it was. Pat got his horse out of the trailer and headed up to join Bill and Donnie. Willie John wheeled his horse around toward me; even a few feet away he had to yell to be heard through the wind. "Get up ahead and get around those. We need to push them up along the fenceline until it jogs to the right and park 'em past there so they don't try to go back."

"Okay."

"Get to going!"

Ahead was a good-sized reservoir, still covered in ice, and the cows had spread out around the sides of it. It meant riding John hard into the wind to cut off half a dozen or so cows at a time and pushing them back along the banks of the reservoir and then up the fenceline on the other side, handing them off to Willie John, then going back for more. I tried taking more the first couple of times, but with just one of me and the cows spaced out the way they were, I couldn't hold them. John responded gamely to my bootheels. He just flat enjoyed chasing cows, and he had done it so much that he was great at anticipating their movements. Which made one of us, and I was grateful for him. It was immensely satisfying, this exhilaration of getting around the cows, of quickness and purpose. Trailing the herd had been pleasant and undemanding, at least back at the drag, but the changing weather and the urgency of getting the cows situated before dark was a challenge.

My feet were frozen; cowboy boots were designed for riding but couldn't touch my lined work boots for comfort. My hands were just as cold. But I still felt damned good forty minutes later when I'd situated the last little cluster of cows far enough up the fenceline, to the place Bill wanted them. Here came the boss now, riding back toward us from the head of the herd, along with Donnie Pettit and Pat Bergan. "Well, how did you like your first trail ride?" My grin was answer enough. "Get your heels down and your toes pointed out so you don't look like a goddamned tourist. It's a dead dude give-away. Thank you. That's better."

We rode back to the trailer five abreast, and I took a sideways look at my four disparate companions: Willie John, big, strong and young—twenty-six—looking almost out of proportion to his short, barrel-chested cow pony; Donnie Pettit, in his late thirties, shifting around in his saddle to share a joke, a smile on his wind-reddened face, as usual; Bill Galt, looking every inch the boss with his Filson duster, Scotch cap and flashy black horse; and Pat Bergan, riding as slow and easy as he sounded, looking quite a bit younger than his sixty-plus years. They all looked happier on horseback than I'd ever seen them otherwise. Trailing these cows was pretty much the first riding job of the year, and it meant the end of the long winter was in sight. With the advent of four-wheelers and other machinery, there weren't that many riding days on the ranch anymore, and they were to be savored.

Ever since the Hard Winter, cowboys in Montana have been bemoaning the loss of horseback jobs. The closing of the range and increased cultivation of hay meant hands had work to do that didn't always involve a horse and a rope. As outfits became more mechanized, that trend only continued. Very few "pure riding" jobs remain, other than for those who choose to wrangle horses for dude outfits. Birch Creek held out for a long time; when Bill worked as a young man on the branding crew, he was ahorseback all day every day. Now Bill keeps a good string of horses, and they are certainly used, but even a hand like Donnie Pettit, who is an excellent horseman, spends most of his time in a pickup or four-wheeler. So this day had a time-honored correctness about it, fulfilling the cyclical motion of cattle to summer range and calving, simple and immutable as the pull of the moon.

* * *

"Nice boots, Pat," Bill said as we approached the cattle guard where the truck was parked. "Where'd you get 'em?"

"Helena," Pat said. "Yup. They feel pretty good." They were insulated, rubberized riding boots, and they looked like a million bucks to me right now.

"Load up the horses and take them up to the corrals," Bill said. "Willie, be careful through Buffalo Canyon. It's pretty slick."

So Willie John and I unhooked the trailer and gave Bill, Pat and Donnie a ride back down to their vehicles at Candy Dan's; then we returned, loaded the horses and headed north. The snow fell silent out of the same slate sky we'd lived under most of the month, and it made the early afternoon sunshine seem like a frivolous dream. The canyon was as advertised, switchbacks snowpacked and icy. Willie John, unfazed, steered us around expertly while he told stories of elk hunting and fishing in this country. We were wending our way along Rock Creek, which I'd already heard was an incredible trout stream, a fact Willie confirmed. A few minutes later we passed a few old buildings on the right-hand side of the road. Painted in foot-tall letters on one wall was the admonition FISHERMAN NOT WANTED. KEEP ON GOING.

"Well, that's certainly straightforward," I said.

Willie laughed. "Courtesy of Pat Bergan," he said. "When he sees fishermen in the creek he's been known to take target practice, sighting his deer rifle over their heads."

How depressing, I thought. This state, this West is already polarized enough: locals versus newcomers, miners and loggers versus environmentalists. Fishermen and cowboys don't need to be at each other's throats. Both fishing and cowpunching have long history in the West, even the supposedly effete, elite fly-fishing. Most of the good fishermen I know grew up local and learned as kids. So you couldn't put this down to simple xenophobia or cultural dissonance. I thought I knew the answer, but I asked Willie anyway. "Why does he feel that way?"

"I think some fishermen down here didn't treat the place very well." Suddenly as an angler I was ashamed. The way some so-called sportsmen treat private land is the outdoor sporting world's dirty little secret. Too many people don't ask permission to hunt or fish, then leave beer bottles and other garbage around, or start grass fires, or take far more than their limit, or drive where they shouldn't, get stuck and demand help getting out, or all of the above. Pat Bergan didn't always hate fishermen.

There's a great deal of bitterness among hunters and fishermen about ranchers and other landowners not allowing access for sport. I've found most ranchers to be pretty good about it. A few are really difficult, and some of them have reason to be. I didn't know how Bill Galt was yet, but I thought it was a shame if a few idiots had spoiled this wonderful stream for conscientious fishermen. I also thought it must be pretty tough to concentrate on your fly presentation with a .300 Weatherby fired across the creek in your direction.

This internal discourse was brought to an abrupt end as the truck lurched to a stop in the middle of paradise. Amazement must have shown on my face as I got out and helped Willie unload the horses. He smiled. "You should see it when everything's green," he said. "It's unbelievable. Take John into the barn and unsaddle him. We'll give them some oats and turn them out."

Willie had parked in front of a massive old barn, about the same vintage as the one near the bunkhouse at Birch Creek. This one, though, was much more intact and was still being used for its intended purpose. I led John in through the big double doors. The old wood of the stalls and poles glowed reddish-brown in the half-light, and the crisp cold air smelled reddish-brown, too—a mix of horses, grain and leather, with a faint but definite tang of urine, equine or perhaps feline. *Eau de old barn,* unmistakable and intoxicating. John snorted steam as I changed out his bridle for a halter, tied him up and unsaddled him.

The door to the little tack room and the surrounding posts were carved with the names of cowboys who'd worked here over the past three decades—those who'd been here before had no time for such foolishness, I imagined—from the taciturn "A.C.—'75" to "Sid," which I knew was my friend Sid Gustafson, Bill's cousin, to the striking "Pat Bergan: 1971–81, 1982 1983 1984 1985 . . ." on up to this year. I thought it quite wonderful that Pat at age sixty should still take pride in carving another year on this door that had swung right through the center of his life for so long.

I found a bag of oats in another little room across from the tack room and poured some in each horse's stall. I found myself besieged as I did so by a loudmouthed, mouse-fattened ginger cat. I found a bag of cat food by the oats and put some out in self-defense. After the horses had eaten their oats we unhaltered them and turned them out into the field behind the barn.

Rock Creek bubbled right through the corrals by the barn. Directly across the road stood a handsome old two-story log and wood-frame house in the Craftsman style, looking very much as it must have in the early 1900s, save the improvement of a shiny new metal roof. House and barn looked about the same vintage. I'd read a little of the history of this country, enough to know that this house was built by Henry Lingshire, foreman and eventual owner of the Walker Land and Livestock Company. "Come on inside," Willie said. "We have to wait for Donnie."

The house was preserved like a fly in amber, pretty much as Henry must have known it. About the only modern concession was a generator for lights and a propane stove in addition to the big Monarch woodstove in the corner of the kitchen. It was a little unkempt; an ancient, tattered deck of cards was still scattered over the kitchen table. A dusty bottle of Johnny Walker Black Label still had half a shot in the bottom, probably owing its survival to the fact it wasn't bourbon, rye or Canadian whiskey, all of which are more popular with the cowboys I have known than scotch. A stack of receipts on a nail over the sink from Edward's Grocery in White Sulphur Springs dated back six years. Pat and June Bergan owned Edward's, after all, so you wouldn't expect to find receipts from Food Farm, the other grocery in town.

A few paperback books were scattered in the dining room, long ago turned into a bedroom: Louis L'Amour, Mack Bolan and the like. A couple of bunk beds squatted awkwardly in the corners. A massive built-in cabinet formed the wall between the kitchen and dining room, opening from both sides. I wondered if Henry and his wife

Mary had hosted dinner parties in this room, with its high ceilings and wainscoting and big front window.

In this rough country, far even now from paved roads or power lines, this house seemed a little incongruous. Seventy-five years ago, to the few people who happened along this way when snow and mud would permit, it must have seemed like Buckingham Palace.

Willie John poked through the larder looking for snoose. No luck. "Place smells musty. Not much food," he said. "Nobody's been here since last fall. Bill's friends, probably, up here hunting and left a mess."

Donnie arrived. We drove to one of the nearby cabins used for hunters in the fall, and helped him load a generator into his pickup. "This is beautiful up here," I said to him, and he replied with a smile, "I used to run the crew down at Birch Creek. Can you see why I do this instead?"

APRIL 1

The yearling heifers have played a mysterious April Fool's joke on us this week, and Tyson and I have not been able to figure it out. Yesterday while we were feeding I found a golf ball, of all things, in the heifer pasture, which is something like finding a piece of asparagus in your coffee. It makes absolutely no sense. Today I found three more.

The heifer pasture is a good deal more than a Tiger Woods drive away from anybody else's property. And nobody at Birch Creek has the inclination or the time to be hitting golf balls.

Nobody human, that is. The image of the heifers surreptitiously practicing their iron shots out here while we're not around, like Gary Larson cows, is hard to suppress.

chapter 8 • thaw

IN TWO DAYS, THE RANCH HAS GONE FROM FROZEN EXPANSE TO sea of mud. There's still a lot of snow in the high country, but water is running everywhere, and everyone on the ranch seems to feel the lift of spring.

I certainly did. My day ahorseback was part of it, I suppose, but I was beginning to have a sense of getting things done. In the Northern Rockies renewal comes with the softening of hillsides and the sibilance of streams freed from ice, with the ratcheting croak of sandhill cranes arriving from the south. It is nothing more than the promise of better days, the reward for getting through, surviving another frozen time.

Also, I was beginning to understand the geography of the ranch. I say beginning, because I had seen only a fraction of the entire spread, and what I was beginning to understand most was how much I didn't know. I thought of the Europeans' pre-Columbian maps of the world, with massively misshapen continents and huge uncharted areas labeled: "Here be monsters."

With the older cows up at Lingshire, feeding was simplified but still took most of the morning. Often we would start with the bulls, who got half a load—five bales. The 120 head of bulls were my favorite cattle on the ranch. They seemed to have more personality than the cows or the yearlings. I thought perhaps it was because of their size—they seemed less frightened of us, more casual.

Every animal on the place was feeling the frisk of spring, but it was most noticeable with the bulls, who looked faintly ludicrous as they lumbered after the retriever in search of breakfast, asses slewed sideways with new exuberance. They made a much larger variety of sounds than the monotonal demands of the yearlings, with my red Angus countertenor noticeable among the bigger, burlier voices. I wondered if the other bulls cast aspersions on his virility.

On the second lap through the bulls, their oblivious disregard of us and the retriever would often become a pain in the ass. They were chowing down on the hay we'd flaked to them on the first feeding circle, and they didn't particularly care that we were hollering at them or beeping the horn, which sounded even more incongruous in the middle of the pasture than the red bull did. Often they would shift their fifteen hundred to two thousand pounds only after getting a front bumper in the ribs. This really pissed Tyson off. "Get out of the way, you assholes!" he would yell. "I broke a headlight on one of them once," he told me, and since both were broken I had no trouble believing it.

Bill had told me that for all their machismo and sangfroid, the bulls were actually more susceptible to cold and lack of feed than any of the other cattle. "Biggest sissies on the ranch," he'd said gruffly. "Make sure you feed them good hay."

Most mornings we fed the other half of that first load to the steer yearlings—male calves who were castrated at branding—while Willie and Jerry took a full load to the heifer yearlings. Then we'd split up the rest: another load for the steers, another half a load for the heifers, four loads for the main herd of older cows, then a load for the pairs—the new moms and their babies, out in the tank field above the calving shed, where I'd first fed with Christian and Fletcher and seen Keith Deal doctor a calf with the scours. Then there were other herds scattered around: the large bunch of cows up on the O'Conner, some newly purchased cows nearer to the calving shed, and about three hundred head that Bill was leasing from local rancher and friend Ronnie Burns, up on the Gurwell. We'd also feed the horses a little hay, and the fats—the steers being fattened for slaughter to provide meat for the ranch.

The hay was scattered in stackyards all over the ranch, fenced areas where the hay was put up in the summer to be used now. The larger stackyards, like the two by the corner near the gate that led to the calving shed and several up on the hill above the shop, were perhaps a hundred yards square and held a dozen or more long rows of bales. The smaller stacks, scattered around the ranch in various pastures, were half as big or less. Some yards held "wild" hay, a mixture of native grasses, and some held hay cut from seeded and irrigated alfalfa.

Certain parts of the ranch still carried the names of previous owners, often homesteaders—the O'Conner, the Manuel up near Lingshire, and the Gurwell, where we were going now. Of the places I'd grown familiar with, the Gurwell was one of my favorites. It was a long, fairly narrow butte running between Birch Creek to the west and the Smith River on the east. It was beautiful and still in the mornings; the views were magnificent. There was a stackyard on the river side and one on the creek side. Often we'd see moose come up from the willows by the river to eat breakfast at the stackyard, and incidentally play havoc with the wooden panels that comprised the stackyard. The panels were twelve feet long and about four feet high, fashioned out of three long rails, two end posts and two crossbars nailed in an inverted V from the bottom corners to the top center. They were wrapped to steel posts and to each other with heavy No. 9 wire. Each morning I wrestled with three wires and undid one panel so we could drive the retriever to the stack. On the north side, a hardened snowdrift had given the moose, as well as who knew how many elk and deer, an easy way up and over the panels, but now we could see they'd managed to smash a couple of top rails in the process. Tyson shook his head. "No use fixing it now. We'll have to rebuild it this summer."

* * *

I was getting much better at the actual work of feeding hay. There was a rhythm to it, a way of doing it so that you were not fighting the retriever—the unforgiving steel of it, chains and tines, teeth and shafts, and its movement, the bumps and stops and starts. There was a way to do it carefully and quickly:

Set the bungee on the steering wheel. Get out of the cab and step onto the tiny lip of the headache rack next to Tyson. Slide the first bale—scoot it forward on each side so there's room for the hay to flake off—cut the strings and take a wrap. Feed it, making sure to free the back corner of each flake so the hay doesn't ball up and jam, which it will anyway if it's frozen. Tie the strings and hang them. Roll the second bale, usually on the second pull; the first is to get in sync with Tyson and to loosen the bale if it's frozen, the second means business. When you roll it be careful of your toes; there's not much room to spare. Climb over the bale and cut the strings on the other side so Tyson can pull them and tie them quicker. Weird thoughts come unbidden: Clambering over the bale never failed to remind me of Jackie Kennedy, the Zapruder film, climbing up onto the back of the Lincoln. Flake the hay off, pulling it toward you to avoid the chains, helping it off with a little kick of the right foot. Economy of motion is important, both for speed and to maintain balance. Next bale the same, and the next. The back bale on the top can be difficult; usually the teeth have grabbed it so that it's hard to roll. Tyson gets down, climbs into the cab and loosens the tines. When the bale starts to slide away from you, yell, "Okay, that's good!" He reappears by your side faster than seems possible, and you roll it. Sometimes it still won't loosen. "Leave it and we'll pull it down with the chains after," Tyson says.

Down to the bottom row of bales. Slide the first. It's the hardest to feed because you have no elbow room at all. Roll the next. Slide the third because if you roll it you'll be squarely in the way of the chains.

The next-to-last bale of the bottom row is usually the easiest on the entire load to feed. If you swing the chains a little, the edges loosen, and then because of its position on the load, the bale flakes off easily. Now, say the back bale on top wouldn't budge. Grab the chain on your side. Stand on the edge of the bed, on the outside of the chain, facing Tyson, who is doing the same thing on his side. It feels like you're going to fall off the truck. You are, but not yet. Your grip on the chain is the only thing keeping you aboard. "One, two, three," Tyson counts. At three, swing your center of gravity, which is to say your ass, outboard, pulling the chain with you. Hard. Presto: the chains pull the teeth, and with them sixteen hundred pounds of hay, down onto the bed. *Now* fall off, being careful to avoid the turning rear wheels. Climb back on, cut the strings, flake the last two bales. Done. A glance at your watch. Good load. Eight and a half minutes, start to finish.

Much more time was spent picking up loads with the retriever and driving back and forth to the pastures than was actually spent feeding—opening and closing gates, dodging potholes and, of course, getting stuck. The mumbled "We need a pull" radio confessions became even more frequent as the mud deepened. The bull pasture, in particular, was a nightmare. Driving through there, even with chains on the four-wheel-drive retriever, was like hitting a downhill putt at Augusta. You're never in control and it's almost impossible to stop.

We started feeding earlier in the morning, trying to beat the mud, and it helped a little. It also gave me the chance to enjoy the sunrise from different places on the ranch. One morning as we headed out of the shop yard to get our first load, I looked back over my shoulder and saw that the first direct rays of sun were touching the stackyard on top of the bar, above the shop. First the top of the haystack, then the whole stack, then, as we rounded the corner, the hill beneath it glowed gold, with everything else still in shade. Spectacular. The spring days were replete with such Kodak moments. Like a kaleidoscope, the ranch scenery seemed to arrange itself into new and endlessly pleasing patterns as the snow retreated, saved from trite beer-commercial prettiness by the rough reality of work to be done in all directions.

On Keith's advice, we usually took a circuitous route to the pairs in order to avoid losing the retriever in the huge ruts going into the tank field. We drove past the gutpile and then up along a ridge, cross-country in the four-wheel-drive retriever, calling the cattle to us. The view to the south was stunning, the Crazy Mountains still snow-covered, tinged with purple from the early light. Usually we'd see cranes flying by to complete the picture. *Flying* somehow seems the wrong word to describe cranes' ungainly, wallowing method of locomotion; they always seem to me to be about to fall out of the air, even though I know they fly as far as a transcontinental jet. They are probably my favorite birds, which is saying a great deal; I am fond of birds of all kinds, but the return of the cranes always means spring here in Montana, and no songbird ever produced a sound I enjoy more than their odd, clattering call.

Tyson certainly has the competence that comes with long experience. Fortunately, he also still possesses the playfulness of youth. By spring we'd grown comfortable feeding together, and certain sporting opportunities presented themselves. One involved dropping hay on white cows or calves. The whiteys are not albinos but are simply part Charolais, a breed of cattle that is big-boned, white and quite rare on Birch Creek Ranch. Bill Galt owns exactly one Charolais bull, a rather obnoxious character named Chuckie, who has a checkered history on the ranch. A few white cows are sprinkled here and there, and very few calves and yearlings. Of the approximately one thousand head of yearlings, about a dozen are white—most of them, I suppose, Chuckie's progeny.

This was rather like life imitating a video game. If a white or nearly white animal (smoke-colored crossbreds counted, but not quite as much) approached the retriever during feeding, great effort was made to land a chunk of hay directly on the animal, preferably the animal's head.

While Tyson clearly has the edge in experience and upper-body strength, which enabled him to aim large flakes with great accuracy, I was blessed with a great deal of luck. If most of the whiteys come to my side of the truck, there's not much he could do, and that's what often seemed to happen. The heifers presented a better opportunity than the steers; by my count there were eight white heifers and only four white steers. I might have been off by one or two, but I don't think so.

The best opportunity usually came on the second feeding circle, because then the truck was traveling through the herd, which was already eating. One memorable morning, we were just starting our second pass when up popped a whitey on the starboard side, where I was feeding. I failed to lead her enough, and instead of getting her on the head, the hay landed squarely on her shoulders. As she tried to get out from under the hay her neighbors were already eating off her back, Tyson muttered, "Only seventy-five points, missed the head."

We'd just started the next bale when—miracle—another whitey appeared on my side. She was farther away from the truck, trying to crowd in but still in the outer ring, which added to the degree of difficulty, but I sailed a square of hay completely over the back of the nearest yearling and scored a perfect hundred-pointer. She calmly shook her head and started chomping, but the evidence was indisputable: A mass of green flakes clung to the curly hair on top of her white head, sort of like parsley on mashed potatoes. "Lucky, lucky." Tyson scowled.

The other diversion we occasionally enjoyed also involved the yearlings. After the last bale of a load had been fed, we would go to one side of the bed or the other and wait until we passed directly by an unsuspecting animal, then jump off and straddle it. A miniature rodeo event ensued, with the startled yearling bucking around in the mud and snow until the rider was dislodged, which usually did not take more than three or four jumps. Great fun, but only if we were ahead of schedule.

Tyson usually turned on the radio in the diesel so we could catch the morning news as we headed to the Gurwell for hay, hopefully sans Rod Stewart. One morning the news was shocking: A suspect in the left-wing loony Unabomber case, one Theodore Kasczinski, had been arrested at his cabin in, of all places, Lincoln, Montana, a little over a hundred miles from here. I shook my head. Montana had been in the news of late for another reason—the right-wing loony Freemen case, playing itself out on an eastern Montana farm, a little farther from us in the other direction.

The West has always attracted oddballs and iconoclasts. It's part of the mythos: Nobody will bother you as long as you don't bother them. Let a man be his own man and stay out of his way. So we attract our share of fanatics anyway. But that doesn't explain all of it. Montana's also attracting movie stars and business moguls by the truckload, or rather the Learjet-load, as well as legions of equity refugees, yuppies who tired of moving up and started moving out instead, looking for a reconnection to open space and "small-town values," and an escape from traffic, pollution, crime, high housing prices and crowded schools. The advent of modem and fax has made it possible for many professionals to pursue their careers without living ten minutes from the San Diego Freeway, or worse. Those social and demographic changes have caused a profound frustration and alienation among many Montanans, who see their homeland turning from a great place to live and work into a virtual theme park full of designer-dressed wannabe Westerners who don't understand what it really takes to make a living on the land.

Here in widely racist, homophobic Montana, where Senator Conrad Burns was accused by a reporter of making a racist remark during his last campaign, only to see his approval rating tick up a point or two, California has been derided for years as "the land of fruits and nuts." And now the fruits and nuts are moving in next door, and a lot of old-time Montanans are angry. The Freemen and other white supremacist, anti-government fanatics have found receptive ears among some of these people, who find themselves marginalized by the state's new economic and demographic realities. Montana, home to the Freemen and the Unabomber, I thought as I fed the yearlings. It's almost enough to make a person flee to Santa Monica. But not quite.

The new season brought an amazing variety of projects that had been waiting only for the change in the weather. One afternoon Keith and I got in the service pickup and headed to the dump up on the bar.

Many items were stored there, wooden fence posts and poles, panels, and various heavy equipment: swathers, balers and several Caterpillar dozers, including a D-8. The "eight," as it turned out, was our destination. Keith needed to remove the big Cat's snowplow blade used in the winter and fit the earth-moving blade in its place. I fetched tools for him and helped him however I could.

The scale of the D-8 was arresting. The blade was easily seven feet high; the heads of the bolts that had to be removed from the huge arms to complete the switch looked to be a couple of inches across. An inch and fifteen-sixteenths, it turned out; I hauled out Keith's big one-inch-drive socket set and found the right socket. Keith added a cheater bar on the end of the socket handle, giving him about five feet of leverage, and got them loosened up. The hydraulic lines were next. "Do you run this much?" I asked him. "Some," he said, "but Bill probably runs it more. He likes it almost as much as the airplanes."

"How much does it weigh?" I asked, still awed at the size of it. I'd seen D-8s before, but never this close up.

"About eighty thousand pounds." He grinned. "Quite a hunk of iron, ain't she? Hand me that spray can next to my toolbox." He carefully sprayed anti-seize compound on all the bolt threads. It took about an hour to finish, and as we left, I thought how commonplace it had become for me to have a "first-ever" experience. It happened almost every day, which for a geezer my age is a delightful thing.

The next day Willie was sent to move some of the Burns cattle from the Gurwell to the Big Field. After lunch, he radioed Bill that one of the calves he was moving had a broken leg. Keith and I grabbed supplies from the calving shed: Ace bandages, stockinette, cast tape, a bucket of hot water, surgical gloves, painkiller, antibiotics. I held the bucket of water in my lap as Keith raced up the bar road. As we pulled up, Bill tried to rope the calf and missed by a hair. The effort spooked the calf and he bolted through a fence into the bull pasture, where Willie managed to rope him. Willie and I held him down while Bill applied the cast at roadside and Keith gave the shots. Then we lifted him into the back of the pickup and held him there while Bill drove up the road to the others. They left me on foot to push the cattle up to the entrance to the Big Field, an aptly named pasture between Birch Creek and Rock Springs. The Big Field was once an entire township—thirty-six sections, or more than 23,000 acres—until Bill fenced off the Dry Gulch, itself an enormous area that stretched north and west from the area around the dump, where Keith and I had been the day before.

I didn't push them too hard, wanting to give little Peg Leg time to get along. He was exhausted before we were halfway there, stopping repeatedly to lie down in the ditch by the road. I'd let him get a few breaths, then get him up and push him, as gently as possible. I wanted to get him to the field, so his mother could find him.

Finally I got the whole bunch, about fifty pairs, into the field—except for one calf who had gotten himself on the wrong side of a fence across the road; his mother, who was following him along on the road; and four other calves who were following her for no good reason other than she looked a little like mom.

Soon Keith came up and together we got the renegade group back down the fenceline and into the field. From the gate, the Big Field swept downgrade forever, out to the horizon, an astounding scope and view. The ranch was like the D-8 Cat: The more I saw of it, the more I was amazed at its scale.

Jerry Huseby certainly earned his keep the next morning. As we fed the yearlings, he nudged me and pointed to one of the steers, a solid black animal that looked like he had been inflated with a bicycle pump. His belly was enormously distended. I radioed Keith, who sent Willie out to get him.

I made lunch that day for the crew, less than elegant, but edible: Two pounds of hamburger browned with garlic and onion I borrowed from my own larder, then simmered with two large cans of tomatoes, mushrooms, basil, oregano, and black pepper, served over two pounds of macaroni. A green salad. Toast and honey for dessert. Coffee.

Then Bill said, "Keith, did you call Doc Schendal to come and do that water-belly steer?"

"Yeah, Bill, but he's not home, he's on his way to Choteau."

"Well, shit, I guess I'd better do it, then." He sighed. "God, I hate this surgery. Hell of a gruesome thing to do to a good steer. No choice, though, he'll die otherwise."

"What causes it?" I asked as we moved into the pulling room.

"Urinary tract blockage—like a kidney stone in people. You'll see here in a minute. Steers are more susceptible to water belly than bull calves, because after castration the urethra doesn't develop the same way. And now, goddamn it, I have to make a heifer out of him." As Willie and I got the steer into the head catch, I couldn't help thinking that this sparkling, polished room would be filthy in a few minutes. Willie showed me how to tie his front legs together and then tie one back leg forward, one back, similar to the branding position. Bill scrubbed up, then pointed to the area of the belly he wanted shaved, and Willie ran the clippers over it. Bill gave the steer a spinal, then a local along the incision path, just as with a C-section. Then he made the cut and dug into the incision with his fingers, searching, searching, swearing, hating what he was having to do. Finally he found what he was looking for and started pulling on it—hard. Yanking might be a better word. It was a fibrous white tube reminiscent of a piece of Romex electrical cable—the urethra. Bill pulled it out to a point where he could get his fingers around it, and sure enough, right where he expected it to be, was a dark, swollen section. "That's it," Bill said. "They always block right where the urethra goes around the sigmoid curve." He snipped the swollen section on one end and the other, a segment about three inches long, and handed it to me. "Want a biology lesson?" he said.

He sewed up the incision around the urethra, supporting the tube and muscle so that a small length of it protruded from the steer's belly. "He'll have to pee through that for the rest of his life," Bill said. Then when he was finished, he cut a small slit in the calf's belly so all the retained fluid could drain away. "Well, that's all we can do," Bill said. "Sometimes they do okay, but he'll never sell for much." I looked at the piece of urethra he handed me, and cut into it with my pocket knife. Out came the stone, about half an inch in diameter. "Pretty interesting, hey?" Bill said. "Get this floor cleaned up."

* * *

All of life is certainly not tra-la-la simply because it's spring. I have a new least-favorite job: washing out bullracks, which are the trailers the Kenworths pull to transport cattle. The bullracks have to be cleaned out after each use; aside from the sanitation and aesthetic issues, cowshit is heavy. In mass quantity, it will make the trailer overweight, which could earn the driver a ticket or even get the rig shut down. As spring progressed, cattle were hauled around frequently, and most of the time, understandably, washout duty fell to me and to Jerry. One of our first efforts still sticks in my mind.

The washout is a bladed area next to the road to the calving shed, right by the creek crossing. The driver backs the trailer up so the back is just a few feet from the creek. Just above the creek, on a flat rock, the unfortunate soul nominated for this job places a five-horsepower pump with two threaded fittings: one for a siphon, which is screwed on and placed into the creek, and one for the hose, which is stored draped over the fence across the road. It's a big canvas hose with brass fittings and a pressure nozzle, like a fire hose. Before putting it on, you prime the pump with a water jug filled from the creek. Then fill the pump with gas and check the oil; then play with the choke and pull the starter cord like a son of a bitch until it starts. Once you've got good pressure in the hose—be careful, there's a leak in the hose, and if you let the leak spray water onto the pump it'll kill the motor—you tote the hose into the back and begin sluicing out the trailer.

A bullrack is quite a complex contraption. There are two main decks, upper and lower, and two smaller compartments, one all the way forward, underneath the top deck and reached by means of a ramp that swings downward; and an even smaller compartment rear of the stairs in the back. The trailers are something of an engineering marvel, with gates and dividers that bolt open or closed, steps and ramps that stow away, and vented sides to allow plenty of air flow. All of those features, though, make the trailers hard to clean. You have to open and close every step, stair, gate and divider, and hose it off top and bottom. The vent holes on the sides are traps for shit, and of course, as any freeway driver who has passed a trailer full of cattle knows, they also allow the refuse to spray out onto the trailer's exterior.

Water pressure is your friend in this task, and the hose and pump will deliver it—if everything is perfect. If the hose gets kinked, if the siphon gets clogged, if the pump isn't working properly, you lose vital psi's that provide the cleaning power. Did I mention you should wear a rain suit? Often I didn't, because we were in a hurry, and so I would get soaked to the skin and covered in watery manure.

Tyson told me about a hand who had contracted giardiasis from washing out a bullrack, and I can certainly see how. With all that water and excrement flying about, it's impossible not to be covered, and if you yell to your helper for more hose, as I did, you're liable to get a mouthful, as I did.

I had barely done the nose and the top deck when the pump ran out of gas, which meant we'd been running it for an hour. Jerry, who was feeding me hose, started to refuel it. I happened to look out right then and yelled, "Jerry! No! That's water, not gas!" He had picked up the wrong jug.

"Did you get any in there?" I asked him. "No, I don't think so," he said. But when he filled it up with gas, the motor flatly refused to fire.

Bill Galt drove by as we were taking turns hauling on the starting cord. "Won't start? What's the problem?" He got out, checked the plug to make sure it was getting spark (it was), fiddled with the choke, took off the air cleaner and tried it that way. Nothing worked. "Did you spray the pump by accident?" he asked. "Seems like something must be wet." By this time I knew what was wrong, but I was not going to rat on Jerry, so I said nothing, just shook my head at his question. "Well, I've got to go. Just keep trying it, and if you can't get it going, call Keith. Why is it off, anyway?"

"Ran out of fuel," I said.

"You should fuel up before you start."

"We did."

"My God, how long have you spent on it? You should be able to finish on half a tank. Washing one of these out is a twenty-minute job."

Well, by the time we drained the watery gas out, refueled and got the pump going again, and got the bottom deck and the exterior done, it was an hour and forty-five minutes, not twenty minutes. We did it faster after that, but I don't think we ever finished in less than an hour.

Tyson tested the wind, throwing a little straw in the air. It was steady, somewhere around 20 mph, I supposed, with a healthy gust every few minutes.

"Ready for a little fun?" he asked me. "Sure," I said, a little uncomfortably. I wasn't sure I liked Tyson's tone, but anything beat washing bullracks, and there was another one to do.

I didn't get much more comfortable when we loaded the maroon pickup with two five-gallon cans of gasoline, a twelve-gauge shotgun and some firefighting equipment: a fire rake and a slapper, which is a truck mud flap affixed to a shovel handle.

"Tyson, you mind me asking exactly what we're up to, here? Have we declared war on somebody? Rod Stewart? The Unabomber? What?"

"Calm down," Tyson said. "We're going to burn the gutpile."

"Are you nuts? Haven't you noticed it's a little breezy out?"

"That's where you come in. We're going to torch it and you're going to make sure it doesn't get out of hand."

"Wonderful."

He followed me to the dump in his own pickup. Up on the hill the wind had freshened quite a bit. I looked sideways at Tyson, and he just laughed. "You can handle it," he said. "Take those gas cans and soak a few of those straw piles and carcasses on the upwind end." A few moments later, he said, "Now comes the fun part. Watch this." He loaded the shotgun with shot shells, which carried explosive charges that detonated on impact. He fired a couple of them into straw piles, and got a wisp of smoke and a puff of straw, but not much else. He frowned. "Put a little more gas on that one there." He fired twice again, and was rewarded with a gout of bright orange flame, black smoke and a crackling sound like a dozen strings of firecrackers.

"That'll get it," he said. "I've got to get back down to the shop to help Willie John. Here's a handheld radio. Call if you have trouble."

Things were okay for a while. I watched as the fire did what it was supposed to do—jump from pile to pile, really moving now, putting up some foul-smelling black smoke, a high thick column of it drifting east with the wind toward White Sulphur.

Not bad duty after all, I thought. Still a wonderful view from up here, snow clinging to the Crazies, and the fire was a beautiful thing, too, in its way.

Yes. Such fatuous maundering came to a quick halt when I noticed a rivulet of flame heading through the grass on the south side of the dump. I sprinted over and slapped it out, but from there I could see four or five more spots where the fire had leapt over the edge of the dump to the grass, closer to the downwind point on the east side. Toward Bill's house.

A solid sheet of flame was advancing through the middle of the dump now. The heat was tremendous. I whacked out the renegade fires and moved around to the north side, where there were a few more.

The wind was gusting now, and as I watched the advancing fire, several tufts of flaming straw whirled into the air and scattered ahead of me. Most of them landed in the dump site; a couple blew into the grass and had to be squelched.

I could see that the real danger would come when the fire reached the east end of the dump. I grabbed the fire rake and hacked a line around the downwind side, coughing and choking in the acrid smoke. I got a line around it the best I could, then looked up to see the baleful orange wall, maybe twenty-five feet away and advancing. The smoke was in my throat, my eyes and nose, and I stumbled smoke-blinded to the north and west until I was in clear air again. I knew I had only a few moments before I would have to go back into the smoke; that leading edge could not be left untended.

For the next hour I went back and forth in a semicircle around the downwind half of the dump, slapping and hacking at the fire. On my second pass across, one tendril of fire almost got away from me because I couldn't see it for the smoke until it had run down into a little gully fifteen yards in front of the main blaze. I pulled my shirt up over my mouth and nose and headed into it, swatting smoldering grass, trying to contain the burn at its edges. There was a bad moment when I was in front of the runaway strip and it curled around and threatened to cut me off, but I finally got it out. I ran back to the main fire on the dead run, worried that it had jumped the perimeter in other places, but it had not, and in another half an hour or so I had a pretty good handle on things. I had to keep checking the runaway areas for rekindles, digging earth up with the fire rake and shoveling it onto smoking spots, but the danger was over.

Just then the radio crackled. "David, do you need any help? Quite a bit of smoke going up there."

"I think I've got it," I croaked. "The wind's still blowing pretty good. I'll wait for a few minutes, then head back."

I knew that I should have called for help earlier, but I had wanted to take care of it by myself. My ego and the west wind had nearly caused a great deal of trouble. As I wearily climbed back into the old Ford and drove back to the shop, I knew it had been a very close thing.

APRIL 8

This Bliss

Feeding five hundred yearlings in warm rain,
spring seems possible. In winter,
even on the gentle days, you knew the chill
was only gathering itself in Alberta
to ride down and freeze your life loveless. Today
is different; you knew it cutting the first bale.
Even as the sea of brockle faces expects,
sandhill cranes fly low to tell you it's real.
The dirty snow of everything that came before
is running water: a sibilant chorus of old lovers
singing for this bliss they could never send you.
You're free to watch deer and Charlie Russell sunsets
and dream of the children of days like this.
You jump happy on a young steer's back and ride,
shocked hocks churning the mud, outrage
soon forgotten with hay. He'll weigh a thousand
pounds by August, and we made it. Our joy
will live through every storm that comes.

chapter 9 • the last heifer

THE HEIFERS WERE ABOUT CALVED OUT BY THE FIRST WEEK IN April. Only about a dozen recalcitrant young cows remained in the pasture behind the barn. The stragglers seemed bad-tempered as a group; one large black heifer with a white face knocked Wylie halfway across the pasture while he was checking the drop one day. She was actively hostile, and we gave her a wide berth. "Ringy bitch," Tyson said with feeling. "I just can't wait until she calves. That's going to be a rodeo."

Another of Tyson's favorites was an older cow, in with the heifers because she had suffered a vaginal prolapse. Tyson had sewed her up under Bill's supervision, and she seemed quite okay, but she was painted with a green P for prolapse. She'd be down the road to market after the calf she was carrying was weaned. Her delivery promised to be interesting.

Frank Grigsby packed up his bedroll and left, his night calving done for the year. We would check the remaining heifers late in the evening and first thing in the morning in addition to the checks during the day.

"I've got a riding job up the road a ways," he told me. "And I go to the Calgary Stampede and work up there every year. I'll probably be back a time or two during branding." I saw so little of him that his departure didn't change my routine that much. I did have the bunkhouse to myself, which was a luxury.

One Saturday afternoon Tyson and I rounded up the few heifers remaining to calve and brought them to the chutes, giving the black whiteface as wide a berth as possible. She snorted ill-temperedly and launched a kick in my direction, which might have connected had her reputation not preceded her.

Cows have an unnervingly deceptive way about them when it comes to kicking. They will seem to be moving *away* from you, often in the direction you desire, damn near out of range, when suddenly they will plant the front end and let fly with a back leg, like a fadeaway jump shot. It is quite comical to look at except when you are the target. In this case, thankfully, the kick did nothing but earn Miss Whiteface a sharp smack across the haunches with a sorting stick. Once we got the heifers in the chute, Bill preg-tested them. This involves running a hand—make that an arm—into the cow to feel for a fetus. Only a couple were "dry," or without calf; the rest were just late.

As Bill tested the last heifer, a large red Angus, No. 235, he said, "Hmm. Yep, she's bred. She'll have a red bull calf."

"Bet," I said.

"What kind of odds will you give me?"

"None. You're the one who made the prediction."

"Yeah, but you must know enough about genetics to know what a long shot that is, both color and sex. I want odds of a hundred to one."

"No way. I'll give you two to one."

He pounced. "If you don't like that bet, how about this? I'll give you a hundred to one and you bet on a red bull calf."

"Okay," I said. "One hundred beers for me if it's a red bull, one beer for you if it's anything else."

"Done."

For a little over three months, more than five hundred of the expectant young ladies had been in the pasture, and they had spent a good deal of their time, naturally, around the manger. Now that the snow was gone, what was left around the manger was deep, oozing and redolent. It was cleanup time, and Bill had both the D-8 Caterpillar and its little brother, the D-6, ready to scrape the, ah, material into a big pile so it could be hauled out.

Only thing was, the big Cats' blades couldn't get right up to the fence, or the edge of the manger, which meant that a more primitive method was required to scoop the stuff out where the Cat could get at it. Namely Tyson and Jerry and me, wading through the goo. Tyson had picked up my old friend the fire rake. It had a big, wide blade on one side that worked well. Jerry and I had shovels, and shovel we did.

As we dug the stuff out, along came Bill, obviously greatly enjoying himself in his first session of the year pulling levers on the Cats, and, I suspected, also enjoying the sight of us flopping around in the muck. He started with the big D-8, then switched to the smaller Cat, finding it more maneuverable in the relatively small space of the weaning lot.

On one pass, he stopped near me and signaled me over to him. I walked over, wondering if my shit-shoveling technique was in for criticism. But no, the boss was feeling rather puckish. After I hopped up on the side of the Cat so I could hear him over the clatter of the big diesel, he shouted into my ear, "It may be cowshit to some, but it's bread and butter to me!"

I noticed as we worked that the few remaining heifers were friskier than usual, running around the pasture, seemingly in good spirits. "Look at them," Bill said to me. "Any animal loves to have its home cleaned. They're happy to have that stuff out of there."

After the shoveling was done, yet another delightful task remained—cleaning the Cats. This involved using shovels and pry bars to extract a great deal of compacted bread and butter from the many nooks and crannies around the Cat tracks.

We finished with that about six, but the evening wasn't complete. The cows in Lingshire still needed to be fed food pellets, called cake, to augment the early-season pasture. Donnie Pettit had driven the cake-dispensing truck into the calving barn, where a truckload of cake, in fifty-pound bags, was stacked on pallets. Tyson, Jerry and I formed a conveyor line, humping the bags off the pallets and up to the hopper at the top of the dispenser, where they were cut and emptied. Fifty-eight bags later, it was full.

I was exhausted, yes. But, I realized, not half as tired as I would have been a month ago. In a little more than two months on the ranch I had lost thirty pounds and four inches from my waist. And it certainly wasn't from dieting. I was eating everything in sight, and not just green vegetables and protein. Chocolate was a precious commodity on the ranch. Everybody was a sweets freak—they were all hooked on the quick energy rush. Often, breakfast for Tyson and Jerry and me while feeding hay was a bag of Milky Way bars. Oh, well, it's not for everyone, this fat farm, but it was working for me. And I realized with a guilty pang just before I lost consciousness that night: *I'm beginning to enjoy this.*

The next day brought another new driving experience, piloting a five-ton dump truck full of shit.

The enormous wall of manure Bill had scooped up in the night lot needed to be disposed of, so while Willie worked on clutch, flywheel and ring gear for Lucy in the shop, Tyson and I tag-teamed the pile. The dump truck held exactly three scoops from the bucket of the loader. When Tyson would dump the third blob of glop into the bed, I'd dump the clutch and head for a borrow pit near the corner of the county road, by the smaller of the two corner stackyards. I very nearly got the truck stuck a couple of times, backing in to dump my load as close to the edge of the pit as possible. Of course, the closest call came just as Bill Galt came driving down the road to check on us. He watched me as I rocked the big truck back and forth, then shut off his truck and started over, either to help or chew me out or both, but just then providentially my rear wheels found some traction. He grinned, waved and got back into his truck, and I sighed with relief. On the next load, though, the truck's tailgate stuck halfway closed as I tried to dump the load, which meant I had to resort to a shovel to clean out the smelly stuff, which meant I began to seriously question my career choice.

The wind had howled all that day, but it reached gale force in the late afternoon, blowing consistently over 50 mph, with gusts far higher. I was waiting for Tyson to finish another load when Jerry came on the radio to ask if anybody knew that the siding was blowing off the bunkhouse.

No, we certainly didn't. Tyson and I went to check it out, and sure enough, a ten-foot flap of sheet metal was peeled back from the south side of the trailer, making it look like a big, ugly sardine can. The wind was pulling hard at the bent metal, threatening to rip it clear loose, and insulation was beginning to blow out. I was about to have a very good view of the bull pasture from my bedroom.

Tyson pulled the four-wheeler up next to the peeled-back exterior wall. "Grab that thing and straighten it out, and try to hold it while I get some sheet metal screws," he screamed at me over the wind, then sprinted for the shop. I got a grip on the metal and bent it back into place, then stepped up on the four-wheeler to get better leverage on it. A gust immediately blew me off and the jagged curl of metal sailed away again.

I wrestled it into position again, my body bowed out like a full spinnaker, and Tyson managed to drive a a few sheet-metal screws. The rest of the screws went in much easier now that we didn't have to support the weight of the metal, and in an hour I had my house screwed back together.

The wind brought us déjà vu. Just in case we had forgotten what six weeks ago was all about, we got to feed hay the next morning in a blizzard. The temperature was in the low twenties, the wind was in the high thirties, and the flakes were the size of quarters. Okay, nickels. Anyway, they were so thick that by nine-thirty we had four inches of snow on the ground. We had to chain up in the bull pasture, and barely avoided the ignominy of a tow. Then, in typically perverse spring fashion, the sun came out by noon, the temperature was in the fifties, and after lunch we worked cattle in half a foot of mud.

Bill had sold most of the yearling steers, and we had to separate the dozen or so young bulls from them and then get them loaded. We moved them down the road and into the weaning lot, then through the alley and onto the trucks. Randy Schmock, top hand at the 71 Ranch, came over with Errol's eighteen-wheeler to help haul.

Not only did I manage to get kicked in the wrist and covered in mud and cowshit, but I also got my first real ass-chewing from the boss. I wasn't alone. He included Tyson and Jerry in his remarks. "You three are moving these cattle like farmers, running them into gates, getting them upset, not giving them a clear path to go. These cattle will be pissed off for a month. Then you deliver twenty-eight to the truck instead of twenty-nine. That's sloppy work, and I'm not going to put up with it. Moving cattle is the most important thing we do here. Only unsuccessful people don't know how many cattle they're moving, or have crews that don't know how many cattle they're moving. I'm not unsuccessful and my crew isn't going to operate that way. Do you understand what I'm saying?"

"Yes, sir," Jerry said. I winced, because I knew what was coming.

"Don't sir me, goddamn it!"

"Yes, Bill."

"And whoever's leaving the locks undone on these corral gates: If I catch you going through one and not flipping the catch, you're fired. And if you have to climb on my gates, don't climb on the side with the latch. It fucks them up."

"Okay, Bill." This in unison.

"Okay." And like that he was over it. He looked at me. "Mr. Goodwrench, get your chin off your chest. You look like shit, by the way. Do you have to get so dirty?"

Tyson told me later, "You know, he probably cares more about moving cattle than anything. He's extremely good at it and he expects all of us to do it right."

"The nerve," I said. "Imagine wanting it done right." I thought that Bill's butt-chewings had maximum impact, for a couple of reasons: One, he knew what he was talking about, and everybody on the ranch knew it. There wasn't a job on the place he couldn't do better than anybody on the crew, with the exception of driving and running heavy equipment. He was no slouch at either, but Keith was nonpareil at those tasks.

Second, he did work hard. I couldn't believe how hard he pushed himself. I can't imagine any other CEO of a multimillion-dollar operation who would take off only a couple of days every two or three years. The ranch was simply his life; he hated to be away. When I took the job, my friend Sid Gustafson, Bill Galt's first cousin, told me, "You won't outwork him. Nobody can."

Just to cap the afternoon, the red heifer calved. Tyson pulled the calf while I was helping Randy wash out his bullrack. He got on the radio to let Bill know. "What did she have?" he demanded.

Well, I got the hard part right—her calf *was* red—but it was a heifer calf, not a bull. Damn. I missed a hundred beers by one lousy chromosome.

Later we took the old maroon truck to the shop, loaded a 200-gallon tank, hose and sprayer on the back, and filled the tank with water.

"What's this for?" I asked Bill.

"We've got to burn some CRP ground at the Stevens."

"Translation?"

"CRP—the Combiners Retirement Program."

"Okay, get serious."

"Cropland Reserve Program. Government program to keep some land out of grain production. Have to burn the grass to keep it in the program."

The next morning Tyson and I loaded the truck with firefighting gear and then took turns yanking the starter cord on the spray motor. Finally, after twenty minutes of choking, priming, hauling and swearing, we got it to fire and ran it awhile to make sure it was warmed up.

Then we hooked up a horse trailer, put three four-wheelers inside and headed for the Stevens to meet Bill and Keith at the CRP field. It had become increasingly obvious to me as the weather changed that if I wanted to be fully useful on the ranch, I would have to learn how to drive a four-wheeler, something I had never done. So after work one day I'd asked Jerry to show me how the throttle, gearshift and brakes worked, and he had. The little buggies had five-speed gearboxes, controlled by a foot-shifter on the left side. The throttle was controlled by the left thumb, and there was both a foot and a hand brake, although Jerry's riding style did not stress the use of these. Then I'd gotten on the red Kawasaki, flipped on the key, hit the starter button and toured it around the shop yard, then up the road onto the bar, down to the antelope sign and back to the shop. No problem. But with the fire, I knew, I'd get my first four-wheeler experience under working conditions.

When we arrived at the Stevens, there was a pretty good wind blowing, exactly the wrong way. Keith and Bill decided to set a backfire at the downwind edge of the field, wide enough so that the main fire, when they set it, couldn't jump it and get out of control. So using a little gas and a butane torch, Keith got the fun started. Tyson and Willie took the truck and left Jerry and me on four-wheelers, and we patrolled the length of the blaze.

The backfire quickly showed signs of backfiring as the wind whipped sparks first over the fence into an adjoining field to the south, then up against the downwind edge of the field. Soon all of us—Bill, Keith, Willie, Tyson, Jerry and I—were running from hot spot to hot spot, slapping and raking at the fire, which reminded me inevitably of my day at the dump.

It took nearly an hour for us to get the backfire controlled. "Stay here and make sure this line holds," Bill said to Jerry and me. "Then you can fan out on the sides with the four-wheelers." Bill and Keith then left for the other side of the field, to light the main fire.

In a few minutes I could see a black smudge in the distance, and in a few more the entire horizon had turned orange. Soon it had grown into a hellish vision, smoke getting higher and thicker, the sheet of flame climbing until it looked like a huge orange tidal wave about to break over us. The crackle of the flame through the dry grass was incredibly loud. The smoke seemed to permeate every molecule of air as it rolled ahead, and the heat made the inferno shimmer and dance before us, like a vision conjured by an evil voodoo priest. The black ribbon of the backfire looked narrower and narrower as the big fire came closer, and I wondered if there had been an enormous miscalculation, and the fire would sweep past the burned strip and be upon us in an instant, but the backfire held and slowly the blaze, starved for fuel, abated into a sullen glow.

By this time Bill and Keith had driven up to see how it was going. "Okay, get going up the road there on the four-wheeler," he told me. "Make sure it doesn't cross the road anywhere."

By the time I got a hundred yards up the field, it already had. In fact, the outlaw branch of the fire had a pretty good head of steam up, racing along a little draw toward some heavy brush. I decided not to be a hero this time and got Tyson on the radio right away. He and I got around it quickly, and a couple of other incursions as well. That took an hour or so, and as we finished, Bill radioed us to meet him at his truck and eat some lunch. Bill had stopped at Edward's and bought cold cuts, bread, mayonnaise and, of course, Pepsi. He inhaled Pepsis, guzzling at least a six-pack a day as he drove around the ranch. I had a thermos of coffee in the maroon truck, and when I brought it out Bill snorted and said to Keith, "I guess he hasn't heard how I feel about coffee."

"And how's that?" I said.

"I don't like coffee breaks."

"How's it different from guzzling Pepsi all day?" I said rather flippantly.

"You can't drink hot coffee on the run," he said. "I've been known to dump thermoses out, haven't I, Deal?"

"Yep," Keith said, and went to his pickup and got his thermos and poured himself a cup of coffee.

"Goddamn it," Bill said, disgusted.

I'm beginning to know every square inch of the shop. I know where everything is: Welding rod. No. 14 metric socket. D-6 service manual. Keith's secret stash of clean grease rags. Carb cleaner. Brake fluid. Impact wrench. Caterpillar-yellow spray paint. U-joints that just might fit an old Ford. Pop rivets. Cable ties. Spade connectors. Sharpening stone. Gravity-defying *Playboy* foldout fraudulently inscribed "To Billy—Thanks for a great time." Hex bolts. Pipe wrench. Bolt cutters. Rat poison. .22 long bullets. The only air nozzle that works on truck tires. Gasket scraper. Tire iron. Torque wrench. Great Falls phone book, 1985. Ninety-weight oil and pump. Whatever it is, I can find it. I may not be the handiest with it once I've got it, but I'm improving.

Working around the ranch, particularly around the shop, you constantly see things that translate to work—things that need doing. Sometimes you can do them right when you see them. More often, whatever you're doing at the time has priority, and you make a mental note: When I have time, and nobody's telling me what to do, there's something I could do. Or you'll think, I bet Keith or Bill is going to put me to doing that pretty soon.

The list never gets shorter. I can only imagine what Keith's and Bill's lists must be like. Today Keith's list and mine coincided: The waste barrels in front of the shop were filled to overflowing. In fact they had overflowed: beside them was a growing collection of scrap metal, worn-out truck parts and just plain garbage. "Get that stuff emptied and hauled off. Burn what you can and take the rest to the dump. Rake the ground in front of the shop and reefer and get every scrap of waste cleaned up."

Then there was Lucy's ruined engine, sitting in a puddle of oil on the shop floor. "Drag that out onto the apron so we can take it to exchange, and cover it with a tarp. Clean up all the oil." That took Jerry and me both. I was careful not to use my back, but the strain caused a shooting pain in my right shoulder, down my arm to the elbow. Still, I'd been eyeing it for two months, knowing it had to be moved, and it felt good to get it done.

Seven-thirty A.M., after feeding, is a weird time to be sorting tires, but that's what Jerry and I were doing. Maybe a dozen tires of various sizes and rap sheets were languishing around the tire machine. We divided them into three groups: good, fix and junk. We tossed the junk outside to await a trip to the dump; mounted the good ones and patched the others, then put them where they belonged, either in the brown shed or as spares on various outfits. One tube-type truck tire needed fixing, and Jerry said, "I'll do it." Willie John came in a few minutes later, took a look and snapped, "What are you doing?"

"Fixing this tire," Jerry said.

"That's off the diesel retriever, isn't it?"

"Yep."

"Where's the tube?"

"I'm converting it to tubeless."

"The hell you are." Willie John glared. "Go get that tube and patch it and get it put back together."

While Jerry labored with that, Keith gave me another task: cutting the ice cleats off the track of a bulldozer. The cleats were raised pieces of steel, about half an inch thick by two inches high, and I had to use an oxyacetylene torch, which I'd never done. Keith showed me how to adjust the gas and make the cuts, then said, "Okay. Get it done."

Another new experience, turning the steel orange-red in a blizzard of sparks, then slicing it away. A simple thing, but it felt great.

I checked the drop and found the last heifer—the ornery white-faced cow—had finally calved. Her little bull calf was very weak. Tyson and I rooted for him to get up on his own, for obvious reasons, but he just couldn't quite do it, so we went out warily with a wheelbarrow. While Tyson distracted the cow like a rodeo clown, I picked up the calf, stuck him in the wheelbarrow and started in. She didn't like it one bit, but she'd had a tough day, too, and after making a couple of runs at us, she settled down and allowed herself to be brought to the barn. The little calf still couldn't get up to suck, so we put her in the head catch and very gingerly managed to get two pints out of her and into the calf.

That did wonders, and in a few minutes he was on his feet and looking for seconds. Which, I must say, brightened the hell out of my evening. There's just nothing like seeing one of the newest creatures on the planet wobbling around on legs pointed in four different directions, with a little milk dribbling down his chin.

MAY 22

As we crossed the fence to the road below the shop, we saw an unfamiliar figure walking toward us on the road. He shouted and waved. We waited for him, and in a few moments he came huffing up to us. "Hello," he said. "My car's stuck in the mud, up the road a way. Could I get some help?"

He looked about as local as a Martian. Just a kid, about twenty, dressed not for a muddy road in central Montana but for a grunge club in Seattle: hands long disappeared in the sleeves of an oversized flannel shirt, very baggy brown-and-black-check pants, mud-covered Doc Martens. Buzz haircut, wispy beard, tiny octagonal specs.

"How long have you been stuck?" Tyson asked.

"Since last night," he said.

"Did you spend the night in your car?"

"No, there was a trailer up there, a bunch of junk piled all around. I slept in there."

Tyson and I both stared at him. That trailer was the property of a family of woodcutters who ran the post-and-pole operation at the dump. He was lucky they were not in residence.

Tyson got Bill on the radio, who listened to the story and asked, "What is he, some kind of nut?"

Gary Welch and Lloyd Poe, who worked for Gary, got his car jerked out of the mud and got him on his way, but not before giving him a good roasting about his foolishness.

The incident emphasized for me how extremely isolated the ranch was, geographically and culturally. Contacts like this one with outsiders would only make it more so. The fellow had read about the Freemen, he said, and he wanted to see what it was all about. If he'd walked in on those loggers, he would have found out, for sure.

chapter 10 • heifers in the mood for love

NOW THAT WE HAD ALL OF THIS YEAR'S HEIFERS CALVED OUT, it was time to get next year's crop with the program. That would be the gaggle of yearling heifers we'd been feeding all winter. It felt weird to be preparing to artificially inseminate these precocious little teen-agers. Babies having babies.

"Cross the creek with the retriever, lead them over, then start feeding on the other side, and drive on in. Flake it thin," Keith said with a glare, "and I mean *thin*. Bill will be watching."

Soon we had a corral full of yearling heifers, and we spent the morning working them through the chute. We inoculated them with a drug to get them in the mood for love—or at least to synchronize their heat cycles so they'd all get in the mood at the same time.

Then we glued rather curious devices known as K-Mor patches on their backs. The patches were pieces of flexible white plastic about six inches long, with a bubble on top containing red dye. They were simple but ingenious devices used to signal when the heifers came into heat. When they started their estrus cycle, other heifers would mount them. This "bulling" was a sure sign of heat, and when it took place, the other cows climbing on the heifer's back would puncture the bubble, and the dye would mark the heifer as ready to be artificially inseminated.

The patches were applied with a gooey, super-sticky cement. My job, in addition to helping bring the heifers up into the chute, was to help Keith and Doug Caltrider, who was a partner with rancher George Berg in an artificial insemination business, put on the patches. I would spread the goo on several patches at a time and keep Keith and Doug well supplied so everything went smoothly.

I tried wearing gloves to protect my hands from the cement, but it slowed me down too much, so I had to work bare-handed. Soon I was covered in cement, which made it tricky to get the patches on the cows instead of on myself, but we got them done.

For several days, we kept the heifers in the weaning lot, feeding them there in the morning, which was a pain—lots of tight turns in the retriever, under Keith's watchful gaze—and waiting for them to come into heat. Each evening we'd run them into the night lot beside the calving shed and sort off the ones with red patches so they could be artificially inseminated, or AI-ed, as they say in the trade. After a few times through this process, the yearlings were hyped up, quick and unpredictable.

The drill was the same each night. We'd run them into the night lot, where they'd huddle at the far end. Then Bill would approach slowly, with me a few paces behind him and on the left side, and Jerry behind me. That gave the heifers he sorted off the main herd a "hole" along the entire right side of the lot. Bill would give the cows just enough room for one or two of them in the front of the milling group to break away, and he would let as many of the same kind through as he could. Sometimes it was only one, sometimes it was two or three. "Red!" he'd yell, meaning heifers with patches indicating they were in heat. "Both red." Or he'd yell, "First two red, cut that black one, she's white!" Then Jerry and I would try to cut the white one back into the middle of the lot while Tyson got the two reds out the gate and into the back alley, where Keith would direct them to the right corral. Then as soon as that was done and the gates were set right, we'd let the white one back. Bill would wait to sort more until we had that done.

Sometimes the cuts we'd have to make would be very close. The rule was, make the cut if you can, but if you're already beaten to the hole, back off. That was probably my biggest mistake, trying to make cuts that weren't possible and jamming up the cattle needlessly, but I was learning.

After Bill got the group winnowed down to about fifty head, we'd move them into the alley by the scale shed, which was more confined and worked better for smaller groups. It was much more demanding, though, because you had to make your move quickly and then get back to allow the cows an escape route. If you got caught in no-man's-land out in the middle of the alley, you were dead. Cows would be confused and unsure of where to go and you could be sure of getting an ass-chewing.

As the person right behind Bill, my job, in addition to making quick cuts if needed, was to hurry the sorted yearlings down the alley, handing them off to the gate men, and then hustle back to my spot. The action was fast and required split-second decision making. Good lateral quickness helped, too. It reminded me of playing linebacker. I saw that it was possible in many instances to control a cow from some distance by small movements, but false movements served only to confuse and excite, both with negative result. You had to let the cow think she was getting away, from you and from the others. You had to give her a path to do that. Conversely, if you wanted to hold her, you had to make sure no path is available.

After the third night of sorting heifers in heat—sounds like an all-girl rock band—I was getting a lot better at it. I made a couple of good tight cuts and tried a couple I shouldn't have, and heard about both. When we were done, Bill allowed, "Well, tonight was a little better. But if you all keep exciting them by making the wrong moves, we won't be able to work them at all by next week."

Doug Caltrider and George Berg already were in the midst of artificially inseminating. I went into the portable barn they used to watch them. It was quite a contraption, able to be towed by a pickup, but set up with everything George and Doug needed to work two cows at a time. We'd jockeyed it into position earlier; it barely fit into the double doors in the chute shed.

George Berg's ranch is right at Fort Logan. The old fort buildings are on his property. Eighteen years ago or so, he started artificially inseminating his own heifers there. "I got sick of buying high-priced bulls that would break their dicks, break their legs, tear up the fences. They were good for only one thing, and sometimes they were no good for that. I started looking for an alternative.

"It worked out so well with my heifers that I started doing my whole herd. Then I started doing some for other ranchers. When I got this barn, which makes it possible for us to do larger herds quickly, I took on a lot more business."

As I watched him work, I was surprised at the way the cows walked into the little stall so willingly. I asked George why they did, and he said, "Amazing, isn't it? You'd think you'd have to beat each one of them to death to get them in there, but you don't. I guess it's dark and quiet, and they can see just a little light at the bottom to show them where outside is. Hell, I don't know why. They just always do. Say, do you want to put your arm into one of these, see what it is we do?"

"Okay."

I put on a glove on my left hand, dipped it in lubricant, and reached up through the rectum, almost up to my shoulder, until I could feel the cervix, slightly larger than my thumb, maybe the size of a carrot. This procedure seemed to disturb the heifer not at all. "Feel it? Okay, now if you were going to inseminate her, you'd take this half-cc spiral syringe with your right hand and insert it into the vagina. You'd use your left hand there to guide it through the gaps in the cervix, directly into the uterus," Berg said, "like this." I moved over and he performed the rather startling two-handed operation quickly. Again, the heifer didn't even blink. He triggered the catch and she walked calmly out into the sunshine.

"Only half a cc?"

"Yep, that's enough."

"What semen are you using?" I asked him.

"This is from a bull called Conveyor. We're also using some from Traveler 722."

"As in traveling salesman, I guess. What makes these bulls so special?"

"They're proven to produce light-birth-weight, short-gestation calves with good survivability, good growth figures."

"Why short gestation?"

"That's how you get low birth weights."

"So in effect, they're preemies."

"Well, no, they're normal, fully developed calves. The usual gestation period for Angus cattle is 281 days. With this semen you'll see calves in about 270 days, with a lot fewer C-sections and other birthing difficulties."

"How long did it take you to get the hang of doing that?" I watched him as he finished with the third heifer since we started talking.

"Not that long. It takes a few times to get it down. Of course I've done thousands of them by now."

There was a chance of slightly better than 75 percent that she would deliver a calf fathered by Conveyor with an assist from George. If the pregnancy didn't occur, she would be bred by a "clean-up bull" from Bill's herd when she came back into estrus.

That was our next chore—to get the bulls in shape to do that "only one thing" they do. Only the replacement heifers would be AI-ed. Bill's herd of about 120 bulls would service the other three thousand-plus cows. But first we had to get them ready. On the appointed morning we brought the bulls from the pasture that had caused us so much grief, down the county road to the corrals. Tyson and Keith left me in the road to turn them into the gate next to the fat pen. It was rather a disconcerting feeling, standing in the middle of the road with nothing but your status as a human to induce more than a hundred recently freed spring-frisky jacked-up two-thousand-pound animals to take a sudden right turn. Take one step backward and it's over, I thought. But I held my ground, and they saw me and instinctively sought an escape route. They found one, through the open gate into an attractively green field, and it worked.

That was the easy part. Now we had to work them through the chute. One of the frequent causes of open cows, or cows who have been bred but abort or fail to conceive, is a bovine venereal disease called trichomoniasis—"trick" to most ranchers—and it is a nasty trick indeed. Once introduced, it can spread through a herd quickly; infected cows will abort, usually in one or two months, then go into a heat cycle again, and spread the evil little protozoan to other bulls. So these bulls needed to be vaccinated, but vaccination, as a rule, reduces the incidence of the disease by only a half. So, as an added treatment, we had to flush bull penises. This task involves a bucket of Nolvasan disinfectant and water, a pump much like a bicycle pump, a flexible hose and a very irritated bull. The hose is inserted between the sheath and the shaft of the penis, and three or four good pumps later, you're done. It takes one person to hold the hose—Keith, most of the time—and another to do the pumping. Keith's son Doran helped his dad, doing most of the pumping. The rest of us vaccinated, ran gates and ran bulls into the chute. When they were simply neighbors and breakfast customers in the bull pasture, I enjoyed the bulls greatly. They had a lot of personality. But it's amazing how quickly they can lose their charming nature when you're attempting to shove them through a chute, give them shots and mess with their private parts.

They were a different story entirely from the few head of yearling bulls we'd set aside to work in a day or two. These boys were big. And quick. And they could kill you.

Exhibit A was a cranky Hereford bull, one of only a few in the herd, with horns so wide they interfered with the head catch when Keith tried to slam it closed on him. He hooked his way out of it and ran through the chute and into the "sick pen," with the bulls that had already been treated.

We had to get him back around and into the chute again, but he was having none of it. He was officially angry now, "on the hook." We let him stand and glare balefully at us for about half an hour, and then I figured he was enough over it and tried to get him into the alley and back into the chute.

I figured wrong. He took a pass at Jerry and me, then ran out the open gate into the alley, got up a head of steam and busted right through a closed gate and into another pen. He was *pissed.* So was Keith. "Jesus, what were you thinking? Don't you know he can kill you? Never mess with a bull that's on the fight like that. Let him alone. Now he's broken a gate and God knows how we'll get him back through."

We finished the rest of the bulls. Keith, still angry, set the gates in the alley to open into the chute, then walked through the broken gate and into the pen with the bull. "Just stay out of the way," he yelled at the rest of us.

The bull swung his head from side to side, pawed the ground and bluffed a charge at Keith. Keith stood his ground and yelled at him. "Go on, you son of a bitch!" Still swinging his head irritably, he trotted out the gate and into the chute. We closed the gate behind him and this time Keith maneuvered the head catch around his horns and got him, then gave him the treatment. He swung his head around, trying to gore Keith and me as I vaccinated him. "Now watch out," Keith said, and released him. The bull shot out of there and rammed straight into the sick-pen fence, which shivered but held. Keith made a note of the bull's ear tag number. "That son of a bitch is going down the road," Keith said. "We don't need this shit."

Willie John and I went over and bent the broken gate back so that it would close, sort of, but it would have to be replaced.

Bill Galt's bulls had had their yearly checkup, and they were ready to perform. Everybody has to do something for a living.

MAY 27

Willie John and I had an interesting task this morning. Two bulls culled from the main bunch last week because they had broken dicks—a relatively common bull malady caused by enthusiasm and an inaccurate aim—needed to be rounded up out of the steer pasture and put into the corrals so they could be shipped out.

We got on four-wheelers and rousted the bulls out of the brush by the creek and got them on the way to the corral. I drove ahead to open the gate. Then, in my hurry to get out the gate so I wouldn't turn the bulls back, I put one wheel of the Kawasaki in the creek, which was running full, and neatly flipped it over into the water. I just managed to avoid getting my leg caught under it. The engine submerged with a great sizzle and hiss of steam. Willie John, shaking his head and laughing, helped me pull it out.

Neither I nor the four-wheeler were damaged, if you don't count my pride, but I did manage to spook the bulls, and we had to start over. They were much harder to get the second time; they knew what we were up to. One of them, a big black Angus, turned and took a run at my four-wheeler, smacking one wheel with his broad, flat head as I tried to get behind him. It took almost an hour to get them in. "What's your next trick?" Willie asked politely when we were finished.

chapter 11 • making good neighbors

WHEN I HIRED ON, EVERYBODY WARNED ME: YOU'LL HATE FIXing fence. But after a few months of blizzards, jugs and bullracks, I was ready for something new, and it didn't take me long to discover that fencing expeditions were always an adventure.

"I want the O'Conner fences gone over and gone over good," Bill Galt said at lunch one spring Saturday. "I want the hundred-dollar job. I don't want Doc Schendal's cattle in the middle of mine. Do whatever you have to, but get it done right."

Tyson and I loaded the maroon pickup with equipment. Ty pointed at a new spool of barbed wire and said, "Take some wire off that and tie it up so it can be carried."

I've tangled with barbed wire a few times, fishing, hunting and hiking, and my win-loss record is not good. Nobody's is. Fishing the Bitterroot River a few years ago, I encountered a barbed-wire fence where one shouldn't have been, within flood level on the riverbed, and managed to open up my hand from the heel to the tip of my big finger. A dozen stitches later, my antipathy toward barbed-wire fences had gone from abstract and theoretical to very goddamned specific.

So I wasn't too surprised when I stripped a healthy length of wire off the spool and it went *spang* and gave me a neat, instantaneous third nostril, just below and in between my original two. "I'm going to forgive you for laughing," I said to Tyson, who was enjoying himself loudly. "You've had a tough day, what with the gate and all." Which quieted him down considerably.

The brakes on the diesel retriever had been deteriorating, and that morning Tyson had pulled it up to the gate into the heifer pasture. As we approached I realized, too late, that Tyson had momentarily forgotten about the condition of the brakes. He had put his foot to the pedal as he would have normally done, and had had time to say, "Oh, shit!" before we smashed through the gate. While Tyson and Wylie built a new gate, Willie and I had repaired the brakes on the diesel.

We got the truck loaded with steel posts; post pounders, which were homemade—driveshafts weighted and welded closed on one end; fence stretchers, relatively simple but ingenious devices that clamp onto wire and enable the user to stretch it tight; two bundles of wire, including my hard-earned one; fencing pliers; large nails and hammer; a chain saw; staples for wooden fence posts, clips for metal ones; a couple of shovels and a pry bar.

Even after we had replaced the fuel pump and shocks, just about any trip in the maroon truck promised to be eventful. The brakes were bad, the throttle stuck occasionally, the transmission still jumped out of third gear, the steering was loose and the front end still seemed to have a mind of its own.

We were about ten minutes into this particular adventure when Tyson hit one of the big ruts on the way to the O'Conner and had the throttle stick at the same moment, which caused the truck to lunge violently to the left just as the road went right. When Tyson took his foot off the gas, hit the brakes and turned the wheel, the truck responded by speeding up and lurching onto its two left wheels. "*You son of a bitch!*" Tyson screamed. I thought the truck was going to roll, but it settled back onto all fours with a thunderous, grinding *whump*. Jerry had been dozing; he awoke in a hurry, and the jagged remnants of the passenger-side door handle dug a small divot out of his thigh. My back felt like someone had hit it with a sledge, but I was happy. We were alive, and the truck had come to a stop in the ditch. Right side up.

We got out and took a look. The only thing wrong seemed to be that one of the new shock absorbers had broken off. Back in the truck. Tyson's mood had blackened considerably, but I think he was also relieved that it hadn't been worse.

Worse came quickly. Negotiating a curve around a little lake by Schendal's, Tyson cut it just a little bit close to a muddy spot and very quickly the rear end was well mired—above the rear axle. It didn't take long to figure that we weren't going anywhere without a pull.

Tyson raised Doc Schendal on the radio. Schendal wasn't home, but he said, "Go ahead and try to start my tractor and pull yourself out with that, but I don't know, I haven't started it in quite a while."

Tyson hiked around the lake toward Schendal's while I waited at the truck with Jerry. I could see the tractor in the distance, and I watched while Tyson made his way to it, tried it and started back. I could see the dejection in the slope of his shoulders on the return trip. Schendal's tractor, it seemed, was dead as a rock.

We raised Willie, who was ahead of us on a four-wheeler, and he said he'd go get a truck and pull the maroon pickup out. We left Jerry to help him and Tyson and I grabbed the fencing gear and got started.

The first part of the fenceline next to Schendal's lay in the heart of a swamp, and predictably, after a winter of heavy snowfalls, it was a mess. Every few yards there was deadfall across what used to be the fence. We'd have to chainsaw through the fallen limbs, then rebuild the fence, resetting posts, stringing wire.

Moose love wet, low-lying areas, and from the amount of scat we saw, moose were definitely close by. A covey of sage grouse flushed as we squelched our way along the fence. Mountain bluebirds were everywhere, tiny blue bullets all the more brilliant against the background of old trees and black water.

Finally we broke out into the open. The fence cornered shortly afterward, and Tyson and I split up. He sent me toward a large hill and headed along the cross fence himself.

I made it up and over that hill, and an even larger one beyond it. The fence was in pretty good shape through here. I found a broken wire and spliced it, replaced a few staples and a couple of clips. Then the fence led me down into a beautiful little glade at the creek bottom. I was fixing a post there when Willie called. "Climb back up that hill, I'm coming to get you."

"I can cover this stretch on the four-wheeler," he said when he got me. "I'm going to take you up toward the timber."

I climbed on the back of his four-wheeler, and he ran me over the next rise, from which I could see not another big hill but a huge fucking mountain, big pines and snowfields.

"Follow the fence. It'll take you right up the face of Tucker. It's two, maybe two and a half miles to the top. I've been up there twice, fencing and bear hunting. Stay in touch on the radio. I'll pick you up later. Good luck." He turned his four-wheeler around. "Oh, if you find any horns up there, I'll buy 'em."

Willie found and sold quite a few horns. Elk, deer, moose and antelope horns, shed each spring, bring good money, mostly because the males of some cultures believe that a little ground-up horn induces rampant tumescence.

But I had more on my mind than horn-hunting as I started up the slope. Like bears waking from hibernation, grouchy and hungry. And my quadriceps, already protesting. I was on the lower reaches of this mountain, by any measure, but the climbing was already strenuous. Thick sagebrush gave way grudgingly to scrub pine, then to larger lodgepoles. The footing was getting a little tricky, too. Pine duff on loose rock, then several granite boulder fields, the big gray rocks slick with snowmelt, frequently sent me looking for hand-holds—not an easy task carrying a stretcher, wire and a bag of staples, clips and pliers.

Always, to my left, was the fence. Its condition was deteriorating as I gained altitude. Much of this fence was very old, and clearly hadn't been gone over for years, so repairs were much more frequent now, and I began to worry about running out of wire. When you try to splice old wire, it frequently breaks instead of bending, which means longer and longer stretches must be replaced. I encountered two breaks where all four strands had been trampled and broken, meaning I had to rewire several feet at a time. The bundle I was carrying was perilously close to exhausted.

The terrain was getting much more difficult. I went through a few iced-over patches, using the sharp end of my pliers as a makeshift ice axe, helping with handholds. I was pretty much on all fours the whole time now, altitude and pitch making twenty or twenty-five paces about all I could do without taking a short rest to get my breath. The wind was blowing up here; it smelled wonderful, of pine needles and moss, thick dark humus and fresh snowmelt. In dense lodgepole forest now, I surprised a bull elk and half a dozen cows, not more than ten paces from me when they finally bolted.

Occasionally I would climb out of a patch of forest and into an open park. The view was staggering from these spots. Looking back down to the east, I could see White Sulphur in the distance, along with a series of four small reservoirs that I suspected were on the Hutterite colony south of the ranch. I could not see any of the other hands. From the radio I knew Tyson was working fence down toward the old O'Conner homestead, Willie was on the fence I had left an hour or so before, and Jerry was somewhere on the other side of this mountain. It was incredibly exhilarating, this soaring above the usual—mountain-climbing with a purpose beyond "because it's there," which was never enough for me.

Serendipity struck: I had exactly two feet of wire left when I came upon a full spool, left hanging over a fence post by a forward-thinking cowboy from years past. Another fifty yards up the mountain, I found two strands down for about eight posts, and from the hair caught in the strands and the profusion of tracks in the snow, it was clear that the elk had decided to make a highway through the fence here. With my new spool of wire I was able to fix it, stout as could be.

So it went for another quarter-mile or so. I had no idea how close I was to the summit because most of the time I could see only a few feet in front of me. Then, suddenly, I came over a crest into a huge snowfield. The snow was crusted on top, from melting and refreezing, and I kept breaking through the crust and sinking up to my thighs, adding an aerobic bonus to uphill travel. The bottom three strands of fence were under snow, so I couldn't do much. Fortunately, the snow was so deep that elk, or cows, for that matter, weren't crossing through here anyway. That, I knew, would change in the next month.

I managed to thrash through the snowy open area and into more timber. The snow and deadfall were both worse here. As I looked above me, the fence suddenly disappeared, and when I reached that point I discovered that it was tied off on a tree and took a left turn for about twenty yards, then started up again.

But shortly after that, the fence disappeared in earnest—underneath an enormous drift probably a hundred yards long. I decided to navigate around the drift to find where the fence emerged. Then I slipped off a log submerged about a foot in the snow and fell in up to my armpits. Even at that point, my footing was not all that solid. It felt springy, and I thought perhaps I was on top of some downed tree branches, but in the process of digging myself out I discovered that I had landed *on* the top strand of the fence, which had apparently jogged again beneath the snow.

Just then Willie John's voice crackled. "David, where are you?" I dug the radio out of the snow—it was clipped to the top of my coveralls—and replied, "I'm most of the way up this mountain. Right now I'm in a big snow bowl, and the fence is under about five feet of drift."

"Okay, that's enough for today. We'll finish up tomorrow. Come on down. Radio me after you get below the logging road and I'll come pick you up."

After I regained my footing, I followed my footprints out of there, leaving the new spool of wire where the fence disappeared, to mark my progress. I soon found that going down wasn't as simple as it sounded, either. After a while I moved away from the fence to find an easier route down, but it didn't work out that way. I kept breaking through the snow, and a couple of pitches were so steep the only way I could negotiate them was sliding down sans dignity on my butt.

An hour later I was on relatively flat ground and happy to hear Willie's four-wheeler coming toward me through the twilight. He delivered me to Tyson and the maroon pickup, which he'd managed to extricate, and by shortly after dark I was home, as tired and as pleased with a day's work as I'd ever been.

The next day I hiked up the mountain again and found a few more giant drifts before taking another circuitous stroll out of the snow. When I got back to earth, I found Willie raging at Tyson. A row of wooden fence posts near the O'Conner stackyard were loose. Tyson said he thought they were okay when he went over the fence the night before; Willie accused him of bad judgment and laziness. "Look at that, a new staple in that post," he raged. "The post is falling down and you just rehung the wire on it."

"I didn't think we had time to do all these, and I thought they could make it another year," Tyson said.

"Shit!" Willie said. "The first time a cow touches this part of the fence and walks through it, I get my ass chewed out. Now go back to the shop and get some steel posts. We'll stay up here until we get this fixed."

We replaced thirty fence posts. Actually, it was pretty good practice for me despite all the negative vibes.

Here's the drill: Go along the fence, wiggling each post. If it's very loose or, like some of these, rotted off at ground level, dig out all the staples, freeing all four strands of wire. Take a new steel post, place it, slip the pounder over the top of it, and whang on it. Steel posts have fins on them, chevron-shaped, about a foot and a half from the bottom. Once you get those fins in the ground it's set. Sometimes that takes half a dozen tries with the pounder; sometimes, in rocky soil, it takes fifty or sixty. This fenceline happened to be in extremely rocky soil. Often you'd give a post a dozen shots with the pounder and not gain an inch because you were squarely on a rock. Take the post out, move it over to miss the rock, and start over. Once it's set, clip the strands and tighten them if necessary. Then move ten yards down the fence and do it again.

We didn't get done until dark, and well into the next day, Tyson and Willie were both a little growly. Both of them were good hands, and both felt wronged in this situation. Keith was hauling cattle, so Willie was in charge in his absence, and he took it seriously. Too seriously, Tyson thought. "What's he trying to prove?" he muttered. On the other hand, Willie knew he would take responsibility for this fencing job, and he didn't want to take Bill and Keith's heat for a mistake he didn't make.

Including the Manger ranch, between Birch Creek and Lingshire, that he has just leased, Bill Galt has somewhere around 220 *miles* of fence on his outfit. To build new fence these days costs about $2,000 per mile, so the fences represent a significant improvement—in the real estate sense, if not in the aesthetic sense. They are also very important to the operation of the ranch. When fences fail, stock gets where it shouldn't, often into danger. Bill was concerned, for instance, that his herd not be allowed to mingle with Schendal's because they could be exposed to disease—somewhat ironic, since Schendal is a veterinarian, but a legitimate concern nonetheless. Robert Frost probably didn't have cattle ranches in mind, but good fences certainly do make neighbors happier in Montana.

I've always thought that fences are to the West what a splinter is to your thumb: foreign, painful, unsightly and an unfortunate fact of life. Fences represent limit and compromise and the closing of the range, and therefore many cowboys hate them. When line riders, who rode the rough perimeter of a spread in the old days, became fence riders instead, trading in their six-guns for wire stretchers, it marked the true closing of the frontier, and what once seemed infinite in its possibility now became circumscribed and clearly defined. Fences didn't solve the problem of wintering cattle; to the contrary, in bad storms stock would pile up against fences and become trapped in snowdrifts. But of course the possibilities had never been infinite, and too many cattle had left the once-lush sea of northern grass in ruins. At least fences put paid to the philosophy that there was plenty of grass for all, and they made each rancher's grazing philosophy—and responsibilities—more evident. Since the 1880s, there have been disputes in Montana about ranchers fencing others away from public lands. Now, as access to wilderness shrinks and recreational pressure increases, many sportsmen find fences emblematic of the perceived tyranny of selfish landowners. Access issues will only get worse as the West gets more crowded.

Despite my antipathy to fences and the dire warnings I had received, I was surprised to find that I love fixing fence. It was a way to see remote corners of the ranch I might never have seen otherwise. And there was immediate gratification from it. You could see results. There's no bullshit about a fence. Either it will hold a cow in (or out, which can be just as important if, for instance, your neighbor's bull happens to have venereal disease) or it won't. If it won't, your job is to fix it so it will.

Bill Galt didn't care how long you spent fixing a fence, but it better be fixed when you were done. "If cows get out, he makes you round them up and fix the fence on your own time," Tyson warned me. I especially like fixing old fence. There are frustrations; the old rusty wire often breaks when you're trying to stretch in a splice, or most irritatingly, after you've finished the splice and you loosen the stretcher it breaks where the stretcher was clamped and you have to start over. Often, too, the fence is in sad shape, with posts down. But there is a sense of continuity in fixing up an old fence, a sense of sharing the land with a few of those who came before. I often wondered about the cowboy who strung the fence I was working on. Did he—yes, or she—look at the mountains and amazing lava formations and sagebrush the way I do? Was this outfit good to him? How much did he make on the day he put this fence up? When he went into town to the bar, did he listen to Bob Wills on the jukebox? Hank Snow? Jim Reeves? If so, he was lucky. Did he get married out here? Get drunk? Go broke? All three? My mind didn't tend to take such ruminative flights while I was washing out bullracks.

One afternoon Willie John and I were sent to go around a two-section pasture on the Manger ranch that Bill had just leased. He'd told his friend Ginger Kinsey that she could pasture her registered Angus herd there.

We got to take one of the best pickups on the ranch, a new

maroon Ford F350. The difference between it and its ancestor of the same color was dramatic, and probably amounted to a couple of hundred thousand miles. It ran down the road, straight and quiet. The brakes even worked.

Bill gave us directions: Go past Fort Logan, and where you see a big osprey nest at the top of a pole, where the state highway makes a big turn, keep going straight. Pass a field planted in crested wheat. Go through a gate and you'll be in a hayfield irrigated with a big pivot system. Turn left along the fenceline and you'll run into the pasture that he wanted fenced. "And be careful of that truck!" he said sternly.

As we pulled into the field, Willie gave me instructions. "I'll take the four-wheeler and go to the left. You take the pickup and go straight along this fence. When you get to the corner, head right. Don't cross any fences, stay inside this pasture and you can't go wrong. And don't take any chances with that truck! I don't want any ass-chewings!" He repeated his earlier admonishment, for emphasis: "Shit rolls downhill around here." He said the last with a fearsome glare.

So I stuck the pickup in a creek. Well, damn it, it didn't look like much of a creek—a trickle, really. I got over the first rise, and the field sloped downward, and I could see that at the bottom of the dip was a little spring. I approached it carefully enough. I even drove up a ways along the little creek, where it trickled past a tree, but I didn't see much difference in the way it looked. So I drove back down to where it seemed to get a little shallower, then put it in four-wheel, got up a little momentum and tried to power through.

Bad decision. The front wheels got through fine but the rear end bounced squarely into the middle of the soft spot and dug in, to about the same level Tyson's had a couple of days ago. The rear end was actually high centered, with the wheels turning freely in the mud. Well, it was time to start the shit rolling my way. I picked up the mike, called Willie and confessed. "Don't do anything, just sit there," he snapped. "I'll be right there.

"I told you not to do anything stupid," Willie said. "Why the hell did you do this?" He took a quick look and realized it was hopeless. "I'm going to go see if I can get some help."

In a few minutes I saw his four-wheeler approaching, followed by a new dark-green Ford Explorer. The Explorer pulled up behind me for a moment, then went up the creek about ten yards farther than I had and drove across slowly and easily. Great.

"Pretty easy when you know the country." The speaker was the driver of the Explorer, an angular gray-haired woman. She was Nancy Manger, one of the owners of this ranch, and she was clearly amused at my predicament.

Willie John was not. When I tried to help hook up the tow chain, he grabbed it out of my hands and did it himself, which didn't improve his mood a few minutes later, when as Nancy Manger tried to pull the truck out, the chain broke loose and whipped against the side of her new outfit, creating a noticeable dent.

Willie John and I were both horrified. "That's okay," Nancy Manger said. "We won't tell Bill about that, eh?"

I knew that was a bad idea. If he found out about it and Willie didn't tell him, he would be really incensed. But for the moment, the bigger worry was that the new Ford pickup was still firmly stuck.

"Get your gear and go along that fence," Willie growled. "I'll deal with this."

So I fixed fence and kept an ear cocked at the radio, waiting for the explosion, wondering if I was about to get fired. I got to the corner and made the turn. This side of the fence was much worse than the first side I'd gone over, and I was so busy that I forgot about my screw-up for about an hour.

Then I heard Bill's voice. "Willie, where are you?" and I froze. "I'm on the west side of this fence, Bill."

"Who stuck the truck?"

I keyed my mike before Willie could answer. "I did, Bill, sorry."

"Damn it, I told you to take care of it. That's a new truck. I got it out. Come and get it. And don't take any more chances with it, period." That surprisingly mild rebuke was all he said, probably because Ginger was with him, and he was showing her where her cows would go. He was pasturing them as a favor to her; she was recently widowed, and her job as a Montana highway patrol officer didn't leave her much time to tend to the Angus herd she'd developed with her husband.

I walked back across the pasture and got back in the truck, which seemed no worse for my stupidity other than the telltale layer of mud on the rear end.

I treated the truck like a carton of eggs the rest of the day. I got out and walked the last stretch I had to cover because I didn't want to drive it down into a rough coulee. Finally Willie John and I met, and he loaded his four-wheeler in the back without a word and took the wheel. Apparently, he was much relieved at not getting chewed on himself, and he was pleasant enough on the drive home.

The next day Bill came into the shop just before lunch. I was greasing the D-8, the backhoe and two of the semitrailers, a low flat bed appropriately called a lowboy, and the fancy new machinery trailer. I had a chance to approach him, and I said, "Bill, I'm really sorry about sticking your truck. I won't do it again."

"Well, when I doubly caution somebody and they do it anyway, it pisses me off."

"I understand that."

"Always get out and walk it if you have any doubt," he said. "Just a few yards upstream, you would have been fine."

"I figured that out. Too late, but I learned something for sure," I said. "Bill, Nancy Manger's vehicle was damaged when—"

"I know. That'll cost me five hundred bucks." He shook his head and walked away.

Later that afternoon, we were working around the shop when Willie took a phone call from Bill. "Okay," he said. He hung up and turned to Wylie and me. "You boys get to go fencing," he said. He looked around the yard. "I guess you can take that," he said, pointing to the oldest truck in the stable, a rarely used old white Ford.

We had to pour gas down the carburetor and jump it to get it started. The truck ran pretty rough, and also had a mysterious charging problem that so far had escaped detection; the battery was usually dead.

When we got it running, Willie drove it over to the gas pump to fill up. He left it idling, got out and said, "Wylie, you drive." Clearly a reference to my fuckup yesterday. "And whatever you do, don't let it die. It won't start again."

The temperature had dropped sharply in the past hour. I ran inside and got a jacket, then loaded the fencing gear on the back of the wheezing truck.

"I want every staple on every post checked," Willie said. "We're out of steel posts, so keep track of where you need 'em. Fix everything else you can."

The sky varied from gray to black toward the west—straight ahead as we headed up the hill from the shop. I walked the fence along the road, checking each post as Wylie drove. The cold wind was howling right in my face.

We got to the stackyard at the fence to the Big Field and made our way around it. The fence was bad here; several posts needed replacing, and three strands of wire were down. Wylie found a level place to park, put the truck in neutral and got out to help me string wire. Suddenly it was snowing, and snowing hard. The wind was pushing the snow horizonal, and in an astonishingly short time drifts were piling up against the fence.

We'd gone another half a mile or so without incident, fixing a few spots in the fence and noting the location of half a dozen more bad posts, when Wylie got back into the truck to drive over the next hill, put it in gear and slipped his foot off the clutch, killing the engine. He looked at me, eyes wide. "Holy shit," he said. "Come on, baby." He tried to crank the engine, but the battery didn't even give us an answering buzz.

I got out and fiddled with the battery terminals, hoping that one was simply loose, but it was no use. We were stranded.

I called Willie, then Tyson on the radio, and, oddly, got no response. Finally I said to Wylie, "Let's wait it out and see if it gets better. I'm going to walk up and do some fencing while we're here. If anybody calls on the radio, give me a yell or come get me." I put my head down and headed into the storm again, figuring I could navigate because I had the fence to guide me. The visibility was near zero and the wind was probably steady at 35 mph.

In about half an hour, I figured I'd better get back to the vehicle. The snow was worsening, if anything, and I thought perhaps Wylie might have raised somebody on the radio.

Still nothing, Wylie reported. We were both getting pretty chilled. Wylie had only a hooded sweatshirt for warmth. I figured I'd better start walking back to the ranch for help, but just then, we heard the sound of a motor. In a few minutes Tyson came into view down the fenceline in his pickup.

"Didn't you hear us calling you?" he asked, and I said, "No, we sure didn't. Could you hear me?"

"Yep, we could hear you fine. Something must be wrong with that radio. Let me give you a jump." He radioed Bill and said, "I've got 'em. Their radio isn't working and the truck was dead."

We got the truck going again. I said, "You want us to keep going?" He shrugged. "I don't know," he said. "I guess so."

So I got on the radio and said, "Willie, Tyson got us jumped. We are continuing around this fence, but be advised our radio is not receiving and the weather's still pretty dirty out here."

A few minutes later, Tyson overtook us and waved us to a stop. He said, "Follow me back to the shop. Bill says you're done fencing for the day. I can't believe this weather, and I can't believe Willie sent you out here in this outfit."

Frankly, neither could I. Fencing with no fence posts in a blizzard, in a truck with a bad battery and half a radio, seemed like a dubious proposition. But no harm done. Just some very cold ears and an adventure to talk about. When we got back to the shop, Wylie and I just looked at each other and grinned.

Afternoon, a couple of weeks later. Willie John, Jerry and I took off to fence a pasture called the Nineteen, down on Birch Creek. Willie, armed with a chain saw, headed for the brush down by the creek and left Jerry and me to do the easy part first: two sides of the fence running through open pasture. The fence bore the signs of heavy deer, elk and moose traffic: hair in the wire, top strand loose all along.

We drove a few steel posts, stretched some wire and met Willie where the trees started. He's what you'd call handy. We'd heard the drone of the chain saw; he'd removed deadfall and fashioned a crude but effective brace post from it. He sent Jerry with the four-wheeler up onto the high ground, on the other side of the pasture.

"Now you and I can work the brush without that little fucker in our way," he said to me, which was unkind. Jerry was actually good at fixing fence; he just seemed not to be able to follow through and finish things right. There was always something wrong, something he just couldn't quite get to. And he was beginning to bug the hell out of everybody on the crew, Keith in particular, because of his lack of energy. He was frequently late to work; he'd complain of car trouble or of his alarm clock not going off. Most days he'd come roaring up the county road from White Sulphur at five or ten past seven.

He seemed to have only one gear, dead slow. Jerry had grown up on a ranch and he knew a lot more than I did about cattle, about fence, about horses. But none of that helped overcome his lethargy. And he was totally reckless with the four-wheelers. He'd get chewed out almost daily for running them too hard, and I figured it was only a matter of time before he had a wreck.

I used my stretchers and wire to help Willie John finish the chunk of fence he was working on, which ran right down to the point where the brush became impenetrable for a cow or almost any other kind of beast, including us. Not that we didn't try. "Let's bushwhack through here and see where it picks back up," Willie said.

I was carrying stretchers and wire; Willie had the chain saw. And for the next hour and a half, we tortured ourselves in the swamp. We made wider and wider circles through the brush, crossing and recrossing the creek, but we just couldn't find a way through to the section line. Much of the time we were bent almost double to avoid thick growth, at the same time clambering over downed timber.

"Fuck!" This, savagely, from Willie. Ten paces ahead, he had slipped into a mud hole and gone in up to his waist in black water and mud. I helped him out, took three steps and did exactly the same thing. His turn to laugh, my turn to swear.

I was getting scratches all over my arms and legs from wayward branches. And just to make the afternoon perfect, the mosquitoes were murderous.

Finally Willie said, "Screw it, we're heading back to where we started. There's a road through the middle of this pasture. It's the long way around, but we'll go around on that, then hike uphill to the corner and back down the other side."

Easier said than done, of course. It was every bit as difficult to go backward as forward. When we got to the road, though, we found out that it was fortunate we decided to go that way. Three good-sized aspen had fallen over the road in one hundred-yard stretch. Willie cut them into two-foot lengths with the chainsaw and we humped them off the road.

The trees had almost no bark left on them—the work of porcupines. We saw several of them, and Willie growled, "Sure wish I had my .22, as if we need to be carrying anything else through this shit." It took another hour and a half, Willie and I joking about our forced march. "Everybody doesn't have to know about this," Willie said, a little embarrassed. "Remember, we weren't lost. We just didn't know where the fuck we were."

Finally we made it around the swamp, up the bluff to the top of the pasture, along the fence to the corner, then down the last couple of hundred yards to the brush. Still no sign of Jerry.

The sun was below the horizon when we finished. We started trudging back along the fenceline, and finally we saw the four-wheeler in the distance. "Wow," Jerry said when he caught up to us. "I had quite an adventure. I got the four-wheeler stuck in the creek. Didn't think I was going to get it out, but I finally did."

It would have been impossible for Jerry not to confess; the four-wheeler was covered with mud over the fenders. He was clearly worried Willie was going to chew him out, but Willie just said wearily, "We had an interesting afternoon, too. Now let's get the hell out of here." Willie drove, Jerry got on the front and I rode on the back. The little four-wheeler performed valiantly with nearly six hundred pounds of cowhand on it. It wasn't a completely dignified way to leave, but at least we'd gotten around the fence.

* * *

The Manger fences were in terrible repair, and there was plenty of fence to fix around Birch Creek, too, that spring. So whenever we had nothing else too pressing, the stretchers, wire, posts and pliers awaited. "It doesn't have to be pretty," Bill would growl. "But I want it to hold cattle."

I continued to find much to enjoy about it. Fixing a particularly bad stretch of Manger fence that ran through a red-rock coulee, I happened upon a huge cedar post, very old but still solid, in need of a new top wire wrapped around it. The top of the post, the size of a dinner plate, was covered with lichens of many colors—yellow, blaze orange, green, even blue-black. It looked like a millefiori paperweight, and even in this stunning country it was one of the most striking things I'd seen.

Fencing is a solitary struggle, not against the cows you're trying to keep in, not against the wire and wood, but against the inexorable forces of nature and time, and as I worked I couldn't escape thinking of the West that had gone before. A. B. Guthrie, who grew up not far from here, near Choteau, Montana, wrote:

> *Frederick Jackson Turner put 1890 as the end of the frontier. If I had to put an arbitrary close to it, I would say it ended with the introduction of the automobile and the tractor. Cowpunchers became or gave way to mechanics then, except those who resorted to rodeo. The workhorse went out of the picture, together with the teamster, and the working saddle horse didn't work much.*

There's a lot to that, and the movement away from the horse has certainly been speeded by the advent of the little all-terrain vehicles. Fixing fence, I often found myself wishing that I was ahorseback instead of on a four-wheeler. Fifteen years ago, I would have been. It would be harder to pack steel posts and a pounder, I thought, but of course I would have been using wooden posts, digging them in by hand. It would take a damned good horse not to spook while you hit a steel post a hundred licks with a pounder in this rocky soil.

But I wouldn't have to be driving the long way around these deep ravines, listening to the whiny little snarl of this engine, and I could ride out into this sagebrush just for the king-hell pleasure of smelling it under the horse's hooves and feeling the wind against my face. There would be no radio to call me back, and the boss probably wouldn't fly over in an airplane and check my progress, either, as Bill had just done.

Willie and I spent a couple of days going around one of the larger pastures near the Stevens homestead on the Manger. The weather was changing rapidly, warming and drying. It was a delicious luxury to work in shirtsleeves or with just a denim jacket, feeling the sun on my back, instead of bundled up and moving stiffly like some sort of coveralled android. On the second afternoon a meadowlark, feeling the same way I did, burst into song from a fence post fifty yards ahead.

I couldn't believe how quickly the land had changed. The sun was blasting, and there was no shade. The sandy soil looked as though it hadn't seen water in a year, despite the fact that it had been under snow four weeks ago.

Here's a fencing tip: If you're going to forget to bring a canteen, mistakenly thinking you'll be near the truck all day, don't bring a peanut butter sandwich for lunch. By five o'clock my throat felt like it was lined with rubber cement.

Suddenly I heard a high-pitched bleat coming from just over the next hill. I ran up the rise and saw Willie extricating an antelope fawn from the fence. The fawn was scared but unhurt, and Willie walked a few steps up to me with it in his arms. It weighed no more than twenty pounds and was probably only a few days old. Willie gently deposited him across the fence, and as we watched he took ten bounds and rejoined his mother, who had slipped under the fence a few moments before.

It was nearly eight when we finished the fence; the light lingered now, which often meant we worked later. On the way home I asked Willie how his wife, Erin, was feeling; she was about eight months pregnant now. "She's fine, but she's pissed off about me never getting a day off and working so late." He shook his head. "She'll be mad again tonight. It'll be nine by the time I'm home."

I knew that Erin had recently sent Bill a letter protesting the hours and lack of days off. Bill wasn't angry about it; he just told me cheerfully, "I don't need her help running the ranch."

Traditionally, ranch jobs were twelve hours a day, seven days a week. Erin's dad had ranched, and given his hands weekends off. Bill didn't, although he tried to get the hands Sundays off when he could. I also knew that Bill pushed himself as hard as his hands; that he was approachable and understanding, in general, if people needed time off for some special reason; and that he clearly explained the terms of employment to all of his hands up front. Still, it was a grind, and Erin, pregnant and taking care of their four-year-old son, Beau, must be feeling the strain of being in effect a single parent.

For beginning hands, it was clearly a buyer's market. There were so many people wanting ranch work that a rancher could pick and choose, and if you didn't like the working conditions, he could simply tell you to go down the road and find another outfit. But Willie had been with Bill for several years, and he was a local product, handier than most, a very good mechanic, knowledgeable about horses and cows, tested tough and not afraid of hard work.

I understood Willie and Erin's point of view and I understood Bill's, too. What about his foreman, who routinely worked longer days than anyone and also had a wife and children? It was not a simple issue. I could feel the frustration in Willie's voice, and I knew how Bill felt. I just hoped they could figure it out without a parting of the ways.

JUNE 14

I fixed mountain fence in a thunderstorm today. In the early afternoon the sun broke through just as I got to the high corner of the fence, an old post leaning outward like a sailor in the wind, into the midst of a vast clearing full of yellow daisies, surrounded by old-growth lodgepole pines. It was incredible, that moment immediately after a thunderstorm when the world holds its breath. Nothing moved but a few drips of rainwater off pine boughs. Finally a woodpecker broke the silence, looking for a meal somewhere high in one of the enormous old trees. I looked past the wire into a meadow where time meant nothing. The stillness between the woodpecker's knocks could be a second or a hundred years.

It's easy to get depressed and fatalistic about the state of our beleaguered planet, with development, deforestation, overpopulation and greed rampant on every continent. But after this storm, on this mountain, at two-thirty in the afternoon of this day, the world had never been more beautiful, and that's why I love fixing fence.

chapter 12 • never buy a rancher's pickup

"EVERY YEAR," BILL GALT SAID, "THERE'S A DAY WHEN I CAN look around the ranch and say, 'Okay, we made it. The cows are on the grass, they're doing well, and everything is the way it should be.' It's not here yet, but it's coming."

I remember him saying that, wistfully, on a day in early spring, like a little boy looking forward to a ball game that was still a month away. Bill and Freckles, his border collie cowdog extraordinaire, and I were driving around part of the spread Bill had just leased, the Manger place between Birch Creek and Lingshire, putting out salt, and the day could not have been more beautiful—bright blue sky, fluffy white clouds, and impossibly green grass. But we had just moved the cows to Lingshire, and spring was not quite believable yet. The day was like a facade, a stage set of spring with a month of winter and feeding hay behind it.

Bill stopped at the top of a little knob, where there was no grass. "Throw out four here." I chucked four fifty-pound blocks of salt out the back of his pickup while Freckles chased a gopher.

"That's why you should never buy a rancher's pickup second-hand," Bill said. "Salt rusts out the bed. Go get that block over there and bring it closer. You don't want to spread out the salt lick too much. It kills the grass."

The salt and minerals are as essential for cows as they are for humans, and fortunately the cows crave them, so they will find the licks and use them. "About this time of year we'll find out if the crew's been putting enough mineral out all winter," Bill said. "When the green shoots start to come up, cows with mineral deficiency can get a disease called grass tetanus. Kills 'em dead."

"Bill, the grass here sure looks good. Mighty green."

He chuckled. "Yeah, the tree huggers would come by here and say, 'Now that rancher takes good care of his grass.' And they'd be dead wrong. Overgrazed land always greens up before anything else. This ground has been seriously skinned."

"Tree huggers, eh?" I gave him a look. "I've had a fling with a tree or two. I happen to be passing fond of them."

"You look like you've got splinters in your chest." Bill would often toss off a smart remark, but whatever you were talking about at the time would stay in his mind, and after a while he'd come back to it in a more serious way. So I just waited, and pretty soon he said, "I'm not saying some ranchers haven't made mistakes with the land. They have. I think most ranchers use too many chemicals. I hate chemicals, always have. I always thought anything that smelled that toxic shouldn't go onto the land. I use them very sparingly. We spray weeds. You have to or the knapweed and the spurge will eat you up. But we do it very carefully and sparingly. You'll experience that soon. I've also nearly eliminated the use of fertilizers, even on the little cereal grain we grow. I really try to avoid them because I think they're long-term bad.

"But I resent the hell out of somebody coming in here and trying to tell me how to run my ranch." He pointed to the northwest. "You've been to Lingshire. You've seen what it looks like. Ranchers have taken pretty good care of that land for the past hundred years plus. When people who basically live in a sewer or have just moved out of one try to tell me how to take care of this place, I have to take exception. I love my ground as much as anybody else could—more, because I know it better. And if I don't take care of it, if I can't grow grass on it, I'm out of business."

That made me think. Hard. I've seen some bad overgrazing in the West, a lot of it on public lands. I've seen some riparian areas turned into mud holes. But not on this ranch—and none of it amounted to much compared with what I've seen of clear-cutting, logging-caused erosion and siltation, and cyanide heap-leach gold mining, which turns mountains into piles of toxic rubble. Not to mention ten-acre subdivisions. My new boss had a point.

It's easy to be an armchair—or Land Cruiser—environmentalist. As a fisherman and outdoor writer I'd done my share of pontificating about the environmental effects of the cattle business, but a little knowledge tends to make you realize how much you don't know. I'm not sure there will ever be a higher or better use of Montana grassland than as an intact, renewable source of food for people. Better to use this grassland to grow grass than to clear-cut the Amazon rain forest and grow grass.

I was glad that Bill had taken to picking me up at the shop or the calving shed for trips like this, if I wasn't too busy doing something for Keith. It was a good way to see the ranch and to get to know him better. I used the opportunity to fire questions at him about the ranch, and he enjoyed talking about things other than the ranch, which made our discussions a bit disjointed at times.

We'd chat about everything from capital punishment—he was firmly in favor—to love and marriage (cautiously in favor, having been through a divorce three years earlier). A picture had emerged of a man both reasonable and hard-nosed; conservative by most measures, yet incredibly generous on a personal level; driven, hard-working, single-minded—and completely in love with the life he had always known.

The Manger pastures were pretty ragged. Junk was everywhere—old car bodies, knocked-down fence, broken panels, old barrels, cow carcasses, mounds of just plain trash. But the ground itself, despite the overgrazed condition, held promise. There were extensive hayfields and some higher ground toward Lingshire in a part of the Belts called the Dry Range. The Manger place had at one time been one of the largest outfits in Meagher County, supporting large cattle and sheep operations.

It was a big chunk for Bill to bite off, this lease: another twenty thousand acres or so of ground, and nothing easy about it. The Dry Range was rough, arid, mountain terrain; and the other half of the lease clearly needed every kind of maintenance and improvement possible. "I'd actually like to find more grass to lease," Bill said. "Land prices are high around here, but they aren't everywhere in the state. Cattle prices are way down, but they won't be forever, and if you can do it, now's a good time to expand."

This ran dead against the conventional wisdom, but Bill was not a conventional rancher. He had the ability to withstand some tough times in anticipation that things would get better. Many ranchers, I knew, didn't have that sort of staying power.

The Birch Creek Ranch was on the upswing; many were on the decline. "For one thing," Bill said, "you very rarely see a ranch that's passed more than one generation successfully, staying in the family, continuing to operate. Lots of reasons for that—inheritance taxes and a lack of desire on the part of the young people to stay in ranching." I couldn't help wondering, as he said it, what would happen to this ground after Bill Galt was gone.

We couldn't see the whole Manger spread that day. Bill was taking delivery on some registered Angus cattle that belonged to his friend, Montana Highway Patrol officer Ginger Kinsey. Recently widowed, Kinsey needed a place to pasture her herd for a while, and Bill had offered one of the pastures he had just leased.

Bill picked up his radio mike. "Willie, take Jerry and get Dusty, John and Bose loaded up. Meet me at Doggett's corrals, right there by the highway. Keith will unload Ginger's cows there."

When Keith arrived at neighbor Jeff Doggett's corrals, I couldn't believe the cattle that walked out of his bullrack. Bill's cattle certainly weren't scrawny or malformed, but these cattle looked almost like a different species, a sort of bovine master race. They were all solid black, so perfect they looked like they'd been turned on a lathe, out of ebony. They were squarer somehow, blocky without looking fat. Legs, hips, heads were straight and strong. They blinked at the sunlight and seemed to look around and take measure of their surroundings. The calves were like perfect little miniatures of the adults.

I had been hoping to get another little riding job here, taking these on the short trail up to the pasture where I'd stuck the new maroon truck. But old John had Jerry's name on him today, and I was stuck with shuttle duty, bringing Bill's pickup along behind the drag. Bill had a reason for assigning Jerry to ride. When he hired on, Jerry told him, "I can ride anything with hair." He'd already absorbed a good amount of kidding about that statement, but Bill wanted to see how accurate Jerry's boast had been.

Jerry actually wasn't a bad rider. But circumstances conspired to make him regret his words. A few of Ginger's cows took a detour around some of the Doggetts' buildings across the road, and Willie spurred his horse to get around them. "Jesus, what the hell is going on?" Bill bitched, irritated that his hands had not anticipated the move. He jumped aboard Dusty, the big black horse, and shot after Willie John, who was already around the wayward cattle and had them headed back in the right direction.

Jerry hadn't reacted very quickly, and he was still in the middle of the road. But all the sudden movement startled John, and he shied and very nearly bucked Jerry off.

Bill, back in the road, saw about a foot of daylight between Jerry and his saddle, and pulled up next to him. "Anything with hair, eh? That's the horse I put kids on when they come to visit, and he damn near dumped you in the dirt."

"He kinda surprised me," Jerry said with an embarrassed grin. I felt sorry for him. One thing I had learned about Bill Galt was that he had a finely tuned bullshit detector. He was not a mean person, but he wouldn't hesitate to puncture false pride, to show that somebody couldn't do what he said he could. I remembered the first time I met him and he asked me what I knew about ranch work. Now I was grateful for my completely accurate reply: "I don't know shit."

Well, I reflected, after my adventures in the weaning lot and the borrow pit, I can't say that anymore.

On the way back to Birch Creek Bill and I stopped at the Mint Bar, where I paid off my red-calf bet and watched with interest the deference paid Bill Galt. A ranching community is in essence a feudal society, and sitting at the end of the bar, eating chicken gizzards and drinking a Bud Light, was the lord of the manor. Mind you, in a feudal system there are overlords who are respected, and those who are not. White Sulphur Springs is a very tight little community. It is full of good people, but like many Western towns of its size, it is slow to accept outsiders. William Kittredge, who understands this place as well as anyone, wrote: "The old West takes care of its own and mostly despises everyone else with xenophobic glee." For all the time he spent here as a kid, Bill and the rest of the Galts are, in the spectrum of White Sulphur Springs, outsiders. So the respect he gets is not a given. "You earn what you get, here," Bill Galt said to me, and on this afternoon, it was evident that the earning had been accomplished.

Among those who paid their respects were a former hand of Bill's and a former rodeo champion, both of whom clearly spent quite a bit of time right here in the Mint these days; and a handsome, dapper fellow by the name of Rick Ringling. Rick, grandson of Richard Ringling, looked as much the patrician scion of a legendary family as Bill looked like a working rancher. Rick had not been sorting any cattle in the mud lately. He was wearing Wranglers with a sharp dry-cleaner crease, a perfectly starched and ironed candy-striped Western shirt, gleaming boots and a pristine white straw Stetson. He and Bill chatted a little, chivying each other about everything from ranches to romance, but they held each other at arm's length, clearly, one man at the head of an empire in the making and the other at the tail of one made two generations ago.

Bill's neighbor, friend and fellow rancher Dan Hurwitz came by and had a beer. Ruddy-faced and white-haired with a rakish mustache, Hurwitz is quiet and friendly. His Cross H ranch is south of Birch Creek, making him one of Bill's closest neighbors, and the two ranchers often gave each other a helping hand, sharing trucks, equipment, even literally lending a hand—maybe two ranch hands—if the other was in need.

Bill finished his second beer, and when Dan tried to buy him number three, he declined, and we headed to Birch Creek. This is a man who keeps himself on a very tight leash, I thought. "Seeing those guys who drink in there all the time makes me uncomfortable," Bill told me on the way home. "Look at the time they waste. I don't want to be like that."

It is the cowboy, not the stockman, who is the Western mythic hero. The cattleman usually is stereotyped as a faceless businessman or a tyrant; sometimes he's portrayed as good-natured but ineffectual—like, say, Melvyn Douglas in *Hud*.

In fact, the cattle business has required much more of its princes than that. Nelson Story fought weather, wolves and Indians to become the first man to trail cattle from Texas to Montana, then sold cattle to beef-hungry miners, helped to establish the town of Bozeman, fought the Sioux to protect the new settlement, built riverboats to facilitate trade along the Missouri, and became one of the most prosperous and respected citizens of Montana territory. His descendants still ranch along the banks of the Yellowstone River. Conrad Kohrs, one of Montana's first prominent stockmen, had to outrun highwaymen intent on stealing his gold dust when he went on cattle-buying trips. Pierre Wibeaux, son of a French businessman and a pioneer stockman, persevered through the Hard Winter, then bought cattle aggressively afterward, when almost no one wanted them. Before the turn of the century, he had accumulated more than fifty thousand head, and he became one of the most successful ranchers in eastern Montana.

Bill Galt, who had learned well under the no-nonsense reining hand of his father, was clearly of similar mettle. While being a large-scale cattle rancher in the late twentieth century does not require fighting Indians or outrunning bandits, it does require equally long hours, hawklike attention to detail, and financial aggression and savvy. The business is still a gamble, subject to the vicissitudes of cattle prices, land prices, weather, disease, and cowboys. The amount of capital needed to operate is nothing short of staggering.

One day after feeding, Bill radioed me at the shop and said, "Keith says he can spare you for a while. Come on up to the house." Bill Galt dreads the days he has to sit at his desk and work on the indoor half of running a ranch. It didn't take long to see why. His desk was covered with an astonishing heap of letters. He sat down at his computer, called up his Quick Books program, loaded a stack of blank checks in his laser printer and said, "You can help me organize these by category." Other than the occasional notice of a bull sale in Columbus or a magazine advertising used front-end loaders and Caterpillars and such, the entire pile of mail was bills.

"One of the scary things about this business is that 90 percent of your income comes in one or two big chunks every year, when you sell calves or yearlings." He added, with a dry chuckle, "The bills come every day."

"Is this a month's worth?" I asked.

"I wish. This is just what came in this week." Montana Power: $1,680. Meagher Motor: $1,740 (including Lucy's engine and the shock absorbers, fuel pump and odds and ends for the maroon pickup). Montana Tire, $883, *one* tractor tire. AquaTech, $1,158, parts for the irrigation systems on the Manger place. On it went. Bills for almost everything imaginable: $770 for bovine antibiotics; $630 for telephone service, which reminded Bill he hadn't collected from Christian for his long-distance bill; $1,205 for blow-in insulation for the little cabin we built.

"Everybody thinks ranches like this one are covered up with money," Bill says. "They don't realize what it looks like on the expense side. And then when beef prices take a dump, my margin goes down 60 or 70 percent. You think any of these guys will take 70 percent less?

"A ranch about this size, in good shape, is worth somewhere around fifteen or sixteen million dollars, including the improvements. There's probably a little more than a million bucks in machinery on it, and an average of three million dollars' worth of cattle when we're ready to ship. Call it a twenty-million investment. Now it's unusual for me to net as much as a 2 percent return—four hundred thousand. Some years I might make that, but some years when we have a bad winter or the market's bad I might *lose* seven hundred thousand. Which points up the need for having a damn good banker.

"Also it shows you that if I could sell out and take the money—which I can't, since I'd lose a huge chunk in commissions and taxes—I'd make more than twice as much as my good year *every* year if I just put the money in certificates of deposit."

"It makes you wonder why anybody's in the business."

"They're in it because it's a way of life and they love it, just like cowboys have to love working all those hours for low pay. That doesn't mean I don't look for ways to make money. Right now's a downtime in the market and I think it's a great time to buy grass and cows if you can get them right. That's why I leased the Manger place and why I'm out looking for ranches to buy."

"Tell me about the Manger place."

"It's what's left of a ranching empire, I told you that. The surviving members of the family have leased it out for a few years now. Bill Loney used to have it, and then Jim Witt, who used to be the cow boss at the 71. Witt just sold out and bought a piece of ground of his own in North Dakota. As you saw, the place is pretty shabby—Witt didn't have much money and got by as cheap as he could.

"I leased it for less than he was paying because I told them I'd have to do a lot of work to get it functioning right. The dam at the reservoir needs work—you'll see that soon—as well as the big water systems, the pivot and the wheel line. It might work for me because I can hay it efficiently with my crew and equipment, and it's located right between Birch Creek and Lingshire. A bunch of the lease is in the Dry Range, and I'm going to try to run some cattle up there."

A few days later, I spent an hour steam-cleaning the diesel retriever, getting wet and filthy in the process, so it should have been no surprise that Bill drove up and said, "Come with me." We loaded twenty more fifty-pound blocks of salt in the back of Bill's pickup. I managed to get a finger smashed between two of them; the nail turned black immediately and it felt like—well, like it had been smashed.

"As much as I hate the book-work days, these days make up for it," Bill Galt said, "when I get to drive around the ranch, putting out salt, checking fence, checking cattle. Looking at green grass." He was feeling pretty springy, I figured.

We went by some heifers that we'd just turned out to pasture. "Look at those girls," he said. "They look so much better than they did three days ago. Their hair is up, they're content, standing up; they're just in far better shape. They're not made to be in a corral, being treated the way we had to treat them. They're made to be out here on the range and they show it."

He was even happier when we checked on some pairs. "Look, they're perfect! The calves are big. The cows aren't fat, but they're good. Everything is supposed to be going to the calf, and it is. They've obviously got great milk. Lots of people who didn't know better would say these cows are poor, but they're not. They're in what you'd call real good range shape. They're not supposed to be fat right now."

We put out salt on Section 19, the pasture where Willie and I took our hike through the swamp. "This is one of my favorite places on the ranch," Bill said. "Good grass, and I love the trees. Isn't this pretty down here?" We were driving on the road through the middle of the section. "Who cut these aspen?" Bill asked. "Willie and I did," I said. "Well, next time move 'em farther off the road," Bill said. "Hang on a minute," I said, and hopped out and dragged the big rounds farther away so there would be more clearance.

Willie called Bill on the radio. He was looking for a bull that had escaped the other day when they were trailed down to the corrals. He'd found him once, but he'd gotten into a brushy area, so Bill said, "You want me to get in the air and take a look?"

"It would probably be a good idea, Bill," Willie said—which was exactly what Bill wanted to hear. He relishes any excuse to fly. He took me along this time to help look for the bull. He had his Piper Supercub in the hangar near his house, but it was too muddy to take off on the nearby strip, so we drove to the White Sulphur airport, where he had his other airplane—a Cessna 185 Skywagon. The Skywagon is the aviation equivalent of a Jaguar roadster. Much more powerful than its more economical cousin, the more common Cessna 172, the 185 is a hot-rod airplane that also offers comfort for four passengers. It is in fact the most powerful Cessna ever made, and it was a steal at $80,000, which is what Bill paid an Alaskan bush pilot for it. Today it is worth on the order of $140,000. A trim, elegant red-and-white beauty, it makes the rest of Montana quite accessible from the otherwise isolated ranch.

It is not a plane for beginning pilots, and Bill is certainly not that. As I watched his careful preflight and letter-perfect takeoff procedure, I judged him to be very meticulous, a good trait in a pilot who flies in mountain country and uncertain weather. "It's a very demanding airplane. They say there are two kinds of 185 pilots," Bill said with a smile. "Those who have crashed, and those who are going to."

Today, though, engineering and technology was no match for one rogue bull. "Damn it, he has to be lying down in that brush," Bill said. But we flew it for an hour and saw no sign of him.

It was wonderful to see the ranch from the air. I gained a lot of perspective on the countryside. It was good to see the hayfields, with their ditches and lands, as the spaces between the ditches were called, and the herds in the various pastures. I pointed out the right side of the airplane and asked, "What are the brown circles in that pasture?"

"That's where the steers didn't clean up the hay because you fed the flakes too big," he said mildly, and I realized that we were looking at money wasted there on the ground that should have translated into weight gain.

Repeatedly flying over the brush involved a lot of turns, and Bill doesn't believe in wasting fuel, time and airspace making wide, leisurely turns. Instead, he hauled the plane up short like a cutting horse every time.

"You look a little pale," he said after one particularly violent maneuver. Actually, I was feeling a little pale, but I wasn't going to give him the satisfaction of puking. Fortunately, I hadn't had lunch yet.

He said, "You know, in this thing you can experience positive G forces"—he yanked back the stick and forced the plane into a near-vertical climb, pinning me to the seat like an insect on a specimen board—"and negative G forces." He pulled up and over into a steep dive, and my harness was the only thing keeping me from crunching against the overhead.

"That's cute," I croaked, hanging on to the bar at the front of the cockpit and trying to smile casually.

"But you don't really get the full effect until you've fallen off a cliff."

"A cliff? What . . ."

Too late. We were sliding along the top of the Gurwell, the alfalfa practically rubbing the fuselage, and the instant we passed the edge of the bluff he dropped the nose and we followed the contour of the cliff down to the river.

"Holy shit! You're going to—"

"Don't worry. Now, didn't you get a wonderful view there?"

"Great. Look, I have work to do this afternoon that will require me to be in relatively good shape. If they're picking pieces of me up out of the Smith River with tweezers—"

He laughed. "Hey, I took it easy on you. I didn't fly it upside down or anything."

"Gee, thanks."

We landed a few minutes later, and I felt like doing one of those ground-kissing maneuvers the pope is famous for. But as we drove back westward toward Birch Creek, I thought how lucky I'd been to get the look at the ranch I'd just had. And to gain one more insight into the man I was increasingly pleased to call boss.

MAY 28

Sunday, the day before Memorial Day. After feeding we worked yearling bulls in the rain, sorting off the ones Bill wanted to keep, then vaccinating them. Bill seemed happy enough with our work, and generally in a good mood. We got done by nine-thirty, and he said, "We'll take the rest of today off, then work the holiday."

I drove to Livingston, for the first time this year on roads clear of ice. I kept the window down, let the morning rain splash in and watched the blurred gray edges of the storm clouds sink like a screen door closing on the valley between the Crazies and the Bridgers. The soft smell of wet ground unexpectedly evoked a summer of American Legion baseball twenty-five years ago, playing the last half-inning through a cloudburst, churning through the mud and diving toward the first-base line for a screaming grounder that somehow managed to elude me. I saw it in slow motion replay, the right fielder scrambling to cut the ball off as I lay flat on the damp edge of the outfield grass, the first-base coach's hoarse yell to the runner, "Go! Go! Take a turn at two, take a turn!" I saw the throw, the slide, the mud flying, the umpire's arm coming up in a looping arc, the runner and the third-base coach jumping and screaming in rage. I remembered running through the grass to congratulate the right fielder on his game-saving throw, feeling the guilty pleasure of escaping the consequence of failure, missing the ball and somehow winning, anyway.

I stopped at a crossroads just outside Wilsall and got out and ran down a muddy little ranch road in the rain like an eighteen-year-old, laughing because I was still living that same life, remembering the magic of it.

chapter 13 • branding and brain surgery

"WELL, ARE YOU READY FOR YOUR FIRST BRANDING?" BILL asked.

"Sure," I said.

"Good answer," he said, "because ready or not, that's what we're doing this afternoon."

This was just a training-wheels branding, the last of the calves from the heifers. These little guys were only a few days old; the calves born on the range would be much bigger by the time we branded them.

Thirty-two calves awaited us in the barn. Everybody, I discovered, had a job to do. Keith, of course, had a bunch: castrating the bull calves, giving each calf a growth implant in the ear, and slitting the ears of the heifers, for ease of identification.

Willie John gave each calf three vaccinations: pasturella; a seven-way blackleg, which includes a vaccine for enterotoxemia, often called overeating disease, which results in distended bellies; and a shot called IBRBVDPI3, which includes vaccinations for infectious bovine rhinotracheitis, bovine viral diarrhea and parainfluenza.

The boss himself does the branding—a backwards L for the steers, and a /OO for the heifers. Because of its size, the /OO brand requires three separate irons to be applied correctly. Wylie and I—we provided the entertainment. We were the calf wrestlers, aging greenhorn and hip youngster in baggy jeans, looking more like a snowboarder than a cowboy.

It took a couple of calves for me to get the procedure pretty much figured. It was like this: One of the wrestlers went into the jug where the calves were penned, grabbed a calf by a back leg, then dragged the calf out of the jug. The other wrestler grabbed a front leg and they swung the calf to the ground on its left side. It was up to the wrestlers to call out "bull" or "heifer" so the correct brands got applied and the cutter knew if he was going for the ear or the crotch. The wrestler on the back end sat on his ass in the dirt, holding the right leg out straight and pushing the left, underside leg in the other direction, pinning it there with his right foot. Meanwhile the guy on the front end kneels on the calf's neck and takes the right front leg, doubles it up and hauls it up toward the calf's head. The other knee goes in the calf's back. This combination, done correctly, immobilizes the calf. Then the others move in and quickly do their stuff, implants, shots, cuts, brands.

When I was on the front end, my face was about a foot from the branding area, which meant I got a good lungful of hair-and-hide smoke, acrid as a cheap cigar. When I was on the back end, I was glad the ever-efficient Keith was the one wielding both the implant gun and the castrating knife, both of which were uncomfortably close to my own body parts.

After the others were finished, Wylie and I would release the calves. That had to be done in concert; if one wrestler lets go and the other hangs on, it increases the risk of somebody getting hurt. Then we'd move on to the next one.

Because of my inexperience, a couple of calves gave Wylie and me trouble, leading to sneers and snickers from the rest. As we struggled to get a good grip and flip one calf, Bill, standing by at the branding pot, said irritatedly, "Come on, what have you got?" meaning which sex. "More than they can handle, it looks like," Willie John said. But most of them went surprisingly smoothly.

We were finished in just under an hour. "Okay," Bill said. "Good job. Willie John, check back in here in an hour or so, will you, just to make sure we didn't catch anything on fire." He turned to me. "Had your fill of ranching yet?"

"You're not getting rid of me that easily," I replied, and he chuckled.

"These are just babies," he said. "Wait until we brand some real calves."

I had plenty else to do while I waited. It wasn't that long—about six weeks—before the Friday night in May when Keith told us as we knocked off, "Make sure you're around by six-thirty tomorrow morning. We're going to brand at Lingshire."

So I got up an hour early, got my coffee and fed the dogs in time to get to the shop by six. Doran Deal, Keith's twelve-year-old son, was already there with Keith. So were Tyson and Wylie. Even Jerry was there early, and I could tell he was excited. Branding was a lot more his idea of ranch work than cleaning the calving shed. He was resplendent in a big felt Stetson.

I'd been trying to wear my straw Stetson, now that it was spring, instead of the ubiquitous baseball cap, but I spent too much time chasing it in the wind. Caps didn't look so authentic, but they stayed on better. Sid Gustafson had passed along an old ranch saying: Never hire a cowboy who wears a straw hat or rolls his own smokes. Whenever you need him he'll be rolling a smoke or chasing his hat.

Still, I thought perhaps I'd made a mistake, not wearing my felt hat. Branding wasn't like most days on the ranch. It was an event, a tradition, a social occasion. And for a newcomer like me, a rite of passage.

But I didn't figure my wardrobe made much difference. Bill Galt had told me a story about his dad: One day somebody commented to Jack that he had quite a few cowboys on hand for the work that needed to be done, and Jack shook his head, surveyed his crew and said, "Well, I've got a lot of boots and hats anyhow."

For me, this day was important to show Bill and everybody else that I could be useful. Willie John and I left at six-thirty to pick up calf wrestlers in town. We picked up three high school kids, including Wylie's best friend, introduced only as Beezer. A few other high school kids would meet us at Lingshire, as well as college boys Lon and Joe Hansen, sons of Bill's neighbor Elmer Hansen.

It was cool and cloudy as we drove out past the Stevens on the way to the Antelope Creek branding trap, which is past Bill Loney's but not all the way through Buffalo Canyon to the Lingshire buildings. When we arrived, there was little to do for the grunts. The senior members of the branding crew, which included Bill, Donnie Pettit, Pat Bergan, Bill Loney, Frank Grigsby, neighboring rancher Bob Fowlie, and former Birch Creek hand and bronc-riding star Chuck Swanson, mounted horses and went to bring in the pairs. Wistfully, I watched the horses leave. I knew, though, that this was a day thick with protocol, a certain way of doing things that varied little from year to year. New members of the crew didn't get to saddle up with the boss on what was as much a ceremony as a roundup.

Indeed, this fabled ranch ritual is much unchanged over the last century. The cattle business has always attracted its share of hustlers, rustlers and just plain thieves, and branding has endured as the most effective way of establishing ownership and controlling theft. Before barbed wire, certain enterprising cattlemen would take to rounding up cattle earlier than everybody else, snagging every calf in sight. Joseph Kinsey Howard relates pioneer Montana stockman Granville Stuart's tale of one such "sooner," in 1880: "Near our home ranch we discovered one rancher whose cows invariably had twin calves and frequently triplets, while the other range cows in that vicinity were nearly all barren and would persist in hanging around this man's corral, envying his cows their numerous children and bawling and lamenting their own childless state. This state of affairs continued until we were obliged to call around that way and threaten to hang the man if his cows had any more twins."

These days fences make such depredations harder, but not impossible. When Jack Galt took over Birch Creek and the other Rankin ranches he found quite a few cattle with Rankin brands in neighbors' corrals, and moved quickly and nearly as forcefully as Granville Stuart to get things lined out.

Now, despite protests from animal-rights activists, branding remains the only legal way to establish ownership, and no other system seems likely to supplant it any time soon.

I felt a little silly, waiting around to wrestle calves with a bunch of kids less than half my age, particularly when I reflected that they had more experience at what we were about to do than I did, but mostly I was anxious to get started.

It didn't take long to get the cattle; Donnie and Pat had gathered them over the past couple of days and parked them nearby. Within about fifteen minutes I could see two long black lines moving across the hillside like ants heading for a picnic. As they got closer, I could make out colors in the morning sunshine, brick red and white cattle in among the dusty black, the half-dozen horsemen scattered out around the herd.

Before long we could hear the rumble of their hooves, hear the nervous mooing, mothers and calves calling back and forth to each other, hear the riders urging them along. The sun was well up and clear of the clouds now, and it glinted off the brightwork on saddles and bridles, painting the scene in vivid color. Even the dust looked red as it floated up from forty-eight hundred hooves—six hundred cows and their calves.

We opened the back gate to the trap and stood to the side, eight calf wrestlers plus Doran Deal plus another few hands come to help—rancher Ronnie Burns and Willie's parents, Willard and June Bernhardt.

The horsemen funneled the cattle between us and the fence, pushing them into the gate. A couple of calves got around the far end of the line, and both times Donnie Pettit chased them down and headed them back toward the gate. I was the closest hand to the gate, and the very last calf suddenly made a left turn and tried to get between me and the fence. Like a defensive end boxing the corner, I scrambled to get in front of the running calf and tackled it solidly, turned it around and chased it through the gate. It was a good start for my day.

I got the gate closed, and then we worked on peeling the calves away from their mothers. Keith, Donnie and Bill, armed with sorting sticks, got by the gate to the inner corral, where we would brand, and the rest of us got in a circle, cut off a couple of dozen pairs at a time, and closed in on them until we could winnow the calves out, then went back for more. Bill took a rough count as he, Keith and Donnie let the calves into the branding corral and kept the mothers on the outside. In about half an hour we had a corral full of bawling calves and another full of irritated, mooing moms.

Now we could begin in earnest. Keith had set up the branding pot at one end of the corral. It was a tradition that the rancher does the branding himself, and Bill did, with some help from Frank and Keith, who also manned an implant gun. Bill Loney and Donnie Pettit took charge of the inoculating, handing out syringes and bottles of vaccine to half a dozen other hands, including Tyson and Jerry. "That kid is really accurate with a vaccine gun," Bill said of Jerry. "I want him in there. He does as well as I've ever seen anybody do." Jerry looked pretty pleased with that, as he should. The cutters got their knives sharpened up and ready: Pat Bergan, Willard and Willie John. Then the ropers mounted up and got started: June Bernhardt, Bob Fowlie, Chuck Swanson and Ronnie Burns. We had a team of wrestlers for each of them. I started out wrestling with Lon Hansen, which was about as good as a guy could hope for. Lon was maybe six-one and two-twenty, and he sure knew how to wrestle calves. "You want heads or tails?" he asked me. I took tails; for me, the rear end was easier. "Good, I like heads," he said, and with that we ran out and wrestled the first calf of the day, being dragged toward the fire by Ronnie Burns.

"Bull," I yelled as we flipped him, and just like that Bill Galt was there with the backwards L glowing red, and Willie John had knife and nuts in hand. Donnie and Bill Loney both gave shots, Keith stuck an implant in the ear, and Doran came by and sprayed the cut with antiseptic. Lon and I looked at each other, nodded, and let him go. He jumped up and ran off, his morning and the rest of his life dramatically changed. Rib Gustafson, in his crusty-old-veterinarian style, calls it "brain surgery—changes the calf's mind from ass to grass."

The whole thing had taken maybe forty seconds.

We got up and looked for another victim. Here came June, riding a beautiful white horse, dragging a hefty red calf, which we grabbed quickly. "Heifer," I bawled, and perhaps she resented me discussing her privates, because at that moment she got a foot free and kicked me so hard in the solar plexus that I thought I was done breathing for the year. I saw a ghost of a grin on Lon's face as I recaptured her foot and stuck mine right in her posterior. This time Willard came over and slit the ear, Jerry and Donnie gave the shots, and Frank handled the branding. Heifers took a little longer because three brands were needed—the / and then the two Os.

I worked with Lon for an hour, then with Wylie for another, and with Kurt Burns, Ronnie's son, for a few head, then back with Lon for another long stretch. Almost everything that could happen, did. Ronnie Burns didn't immediately realize I'd downed one calf and dragged me about thirty feet. I flipped another nearly on top of the branding pot, getting yelled at for my pains; got stepped on by Ronnie's horse as he went past; and got kicked so many times I lost count. But by the time somebody yelled, "Lunchtime," I'd gained quite a bit of confidence. I found I was able to flip calves pretty easily by grabbing the tail, lifting up and grabbing the roped leg at the same time. And I was getting a lot better at immobilizing them by shoving one leg hard away from me with my boot while pulling just as hard on the other with both arms. It was not unlike a baseball hook slide, pointing the right foot out as you land on your ass. Landing was the constant, of course; I spent most of the morning square on my tailbone in the dirt. Oh, and in something else Ronnie's horse left for me.

I also found out that I loved it. This was the ultimate Nike commercial: Just do it. When a roper was dragging a calf toward you, there was no time to think about anything but getting the calf down, getting the rope loose so the roper could go get another one, and holding the animal steady while everybody else did their jobs. Staying out of the way was pretty important too. I got a couple of needle sticks during the morning because either my leg was in the wrong place or the needles were or both.

By the time that lunch call sounded, we'd done nearly four hundred head, and had just a few more than half that many to go. And what a lunch. Julia Short had been fixing lunch at Bill Galt's brandings for a dozen years, and to say she had it down was an understatement. Half a dozen different kinds of salads and relishes. Swiss steak. Mashed potatoes. Green beans. Fruit compote. Pie and coffee. And as much of all of it as you could hold.

Julia was an important part of the Birch Creek extended family. Her husband, Jack Short, had been a cowboy's cowboy. He'd had his own spread, but he preferred riding for other ranchers, and he was very good at it. He worked for Bill Galt and Bill Loney for years, including some of that time at Birch Creek. He sat a horse and worked cows with the best of them, and when he died suddenly, cutting firewood on his ranch, his loss was felt keenly.

When Bill put in his radio system, he needed a spot in town to put a repeater radio, to extend his range. He asked Julia if she would mind having it in her house and relaying messages from time to time. "I'd love it," she told him. "It would make me feel like I was still part of the ranch." And when Bill needed someone to cook for occasions like branding and shipping, Julia was glad to do it. And after this lunch, I was certainly glad she was doing it, too.

Julia's brothers were Howard and Harold Zehntner, whose ranch had been in one family, under the same name, for more than a century—the longest of any in Meagher County. Harold and Howard were usually present for Bill's brandings. They'd been busy this morning, so they had arrived late. They had both been incredible ropers. Harold was laid up now and didn't rope, but he enjoyed coming to the brandings, standing next to the branding pot, occasionally taking a couple of calf nuts from the bucket, grilling them atop the branding pot and eating them on the spot. Howard, formerly a national roping champion, was still a superlative roper. Pointing to Howard and Swanny, Bill said with understandable pride, "I've got more champion cowboys at my brandings than any other rancher you'll find."

After we helped Julia pack everything away and had one more cup of coffee, it was time to get back to it. Soon the ropers slowed down a bit, having to look harder and harder to find unbranded calves, and by about two o'clock we were done. We released the cows and calves, packed up Keith's truck with all the gear, trailered all the horses—and drove down the road to Bill Loney's place to do it all again. Loney had about five hundred pairs to finish the day for us.

We were lucky with the weather. It rained some, but not enough to keep us from working. It was an active afternoon for me; the Hansens and most of the high school wrestlers had to leave, so Wylie and Jerry and Kurt Burns and I wrestled all the calves. Bill Loney, quiet and courteous as ever, did the branding, and he smiled at me about halfway through and said, "Had your fill yet?" The truth was, I hadn't. I was working at the intersection of Western reality and myth, and it was a pleasure, beside cowboys like Loney, Pettit, Swanson, Burns, Zehntner. And Galt.

Still, when the ropers finally couldn't find any more unbranded calves, it was close to seven o'clock, and it did feel like a day's work had been done. We'd branded about eleven hundred head, and I figured I'd had one end or the other of about a third of those—somewhere between three hundred and four hundred calves.

We got back to Birch Creek, unloaded the branding gear from Keith's truck, including two five-gallon buckets full of calf testicles, and headed for home. The shower felt so good I didn't get out until the hot-water heater could no longer keep up. I had hoofprints in some interesting places.

Several rainy weekends put us behind schedule, but we would have several more big brandings around the ranch before we were finished—including Loney's, somewhere close to four thousand calves. I wrestled mostly, but got to run a vaccine gun for a while, and even cut off a few nuts just to get the feel of it.

We'd wash the nuts thoroughly each night after branding, then leave them refrigerated in cold water overnight. The crew would spend the next morning in the calving barn, sitting on upturned buckets, cleaning nuts—taking each one and cutting the tough outer skin away with a single-edge razor blade. The clean ones were put in Ziploc bags and frozen to await the annual feast at the end of branding. One more thing I never thought I'd do.

* * *

One day in early May, Bill came into the shop and said, "Try to degrease yourself and come with me." We set off through the gate at the antelope sign and into the Big Field. The clouds hovered low on the hills, but the strong afternoon sun drove through in long swirling shafts moved from place to place by the wind, lighting up little areas of pasture, a hillock, a clump of sage, a cow and a calf, for an exalted moment of gold in the gray.

As we approached Thompson Creek, I was very glad I wasn't driving. The creek crossing on the road was washed out. "I don't know," Bill said, put it in four-wheel, stuck it in second gear, picked a spot and tried to ram his way through, sort of like I did at the Manger a few weeks earlier. As we hit the creek, he said, "We're fucked," and sure enough, we made it through the water but mired hopelessly in the mud on the uphill side. The spinning front wheels threw a huge cloud of mud onto the windshield, leaving us in a sightless cocoon, and with that, Bill switched off the engine, grabbed the mike and said, "Keith, are you around?"

"Go ahead, Bill."

"Mr. Goodwrench and I are in the Big Field, by the Thompson Creek crossing, and we're stuck. Can you bring the loader over here?"

"Sure, Bill, but it'll be a few minutes."

"That's okay." He hung the mike up and looked at me. "Never do this."

"I already have."

"Never do this again. Now come on. We can walk from here."

Up the hillside and maybe another couple of hundred yards or so north we came to a fenceline, and adjoining it the ruin of a corral. "Let's see how you do managing a project," Bill said. "This is the Rock Springs branding trap. We're going to be branding the calves from the Dry Gulch and the Big Field here, and it needs to be rebuilt. I want you to make sure it's done right."

We walked through what was left of the corral. "Replace these rotten rails"—he gave one a good kick, and it splintered—"and here, where the wing is, this needs two new braces, one going this way for the wing, one that way for the fence. This corner post needs replacing. Actually you don't need to replace it, just put another post behind it and wire it up with No. 9. All this hogwire along the bottom, make sure you pull it up and wire it so calves can't get out. Fix this gate here. Repair these panels where you can, and bring some more up from the dump and put them along here, where the fence is too low; they can jump it. All the trash and old wood and wire and stuff, put in a pile over here so we can burn it. Got that?"

"I think so." I gulped.

"Okay. You and Jerry can do it, and maybe Willie John can help you. Keith can probably set a couple of the big posts with the auger."

It started to sprinkle as we walked back to the truck, and by the time we got there it was raining hard. Keith showed up in a few minutes, and he and I got the towrope hooked from the loader to the front of Bill's truck. I had the rope looped around the hitch on the back of the loader when Bill called out, "No! Never put a knot in a towrope. That's how they get frayed and break. Just use the hook. That's what it's for." The loader extricated us easily.

Keith stayed at the crossing, moving some dirt to fix the washed-out bridge. The rain eased and the clouds quickly retreated, leaving an evening clean and cool as a new cotton sheet. Bill stopped the truck on the crest of a little hill, got out and surveyed the vastness of the valley that stretched in front of us.

"See those two cows over there?" he asked. "What do you notice about them?"

"There's only a couple of them, and a lot more calves, maybe a dozen."

"Right. They're babysitters."

"Come on, give me a break."

"No, I'm serious. The other cows are farther out, grazing, and these gals are looking after the calves, waiting their turn. People don't realize what great animals cows are."

He motioned with his arm at the expanse of field in front of us. "This is what cow country should look like," he said. "No roads, no utility lines, no fences, no buildings. Just good grass and cows and water and sunshine." He turned back toward the truck, immensely satisfied. "This is the day," he said. "*This is the day*. We made it through another one."

JUNE 27

"You did real good, looked like you've been around cattle all your life." That was pleasant, coming my way from Harold Zehntner between bites of what most places are deep-fried euphemisms, Rocky Mountain oysters, but here are simply calf nuts.

It was the annual testicle festival, the Birch Creek post-branding party. Just about everyone who took part in the brandings had gathered at the Connexion, the steak house on the south side of White Sulphur that was the town's best place to eat. I was used to seeing the other hands just as covered in grease and mud as I was, and it was good to see them slicked up and celebrating. Everyone was gathered around the big horseshoe-shaped bar, and I said my how-do-you-do's: Julia Short and the Zehntners; Donnie and Rose Pettit; Pat and June Bergan; Willie John and Erin, enormously pregnant; Bill and Pam Loney; Nancy Walter; Keith and Kelly Deal; Jerry Huseby, resplendent in his big black Stetson, his fiancée, Shyla Low, on his arm; Frank Grigsby, Bob Fowlie, Ronnie and Kurt Burns, Tyson and Wylie, and of course an avuncular Bill Galt, sipping on a Walker's and Coke and looking rather pleased.

Before too long we moved into the dining room and powered through salad, potatoes and somewhere around three thousand of the aforementioned items, which have a mild but distinctive flavor. I sat next to Harold Zehntner, and that was when he was polite enough to compliment my calf-wrestling technique. I knew better, but I appreciated it anyway. Great night.

Part Three

Laying, Spraying and Other Delights

chapter 14 • water the grass

TRUE TO ITS NAME, BIRCH CREEK RANCH IS WELL-BLESSED WITH water, that most precious of ranching commodities. Without it, land is worse than useless, and the banker is always at your door.

Birch Creek, Little Birch Creek, Butte Creek, Rock Creek, Camas Creek, the Smith River and countless smaller streams all flowed on Bill Galt's land. Now that the watersheds were giving up their winter stores, we needed to be putting the water where it would do the most good—on the lush hayfields of Birch Creek, as well as the Stevens, up on the Manger lease. Two of the Manger fields had big mechanical irrigation systems, a center-pivot type and a wheel-line sprinkler, but everywhere else the water was moved the old-fashioned way, flood irrigation done by hand.

The ever-demanding Keith Deal provided the primer for Jerry and me, the crew's only two irrigation neophytes. Birch Creek had two kinds of hayfields—those seeded in alfalfa, and those that produced "wild hay" from a mixture of naturally occurring grasses. Our job this morning was to open the water into several fields of wild hay. The wild hay was nowhere near as rich in food value as alfalfa but was useful nonetheless. "It's just a little better than straw, as far as nutrition goes, but it all looks good against a white background," Bill had told me.

Keith drove up onto the bar, then took a left and headed for the bottomland down by Little Birch Creek. "Your wild hay, you can leave the water running on it for a week or so before you have to move it," Keith said as we bounced along. "But the alfalfa is different. If you flood a section of alfalfa for more than twenty-four hours, you'll kill it."

He parked the truck just off the road and walked in toward the creek. We were in the irrigator's uniform: hip boots and a shovel across the shoulder. Spring was heavy around us: the nearing rush of snowmelt, the thickening green of grass and cottonwoods, the call of mountain meadowlarks, and suddenly, to our left, the stir of something very large in the brush, which turned out to be a yearling moose, breaking large boughs as he rumbled through a thicket, fortunately away from us.

We came to the creek, swollen and quick, and walked through it. The feel of fast water around my feet made me think of fishing, and I thought how far away I was from other springs, when I would try, on days like these, to leave my indoor work behind and fish a river before the melt turned it into a torrent. This new outdoor work would not allow such frivolity. Work came first, and after work there was only eating and sleeping and more work. Yet I felt none of the frustration I would have felt working at a desk and not being able to get away. I missed fishing, certainly; but in this work there was joy beyond any other, being out-of-doors with a purpose at seven A.M. and back inside twelve hours later with a feeling of weary accomplishment.

This irrigation, though, was going to be difficult. There was no compromise to it; either the water went the way you wanted it to or it didn't. I watched Keith wrestle with a primitive headgate, several boards a couple of feet long, slid into grooved channels. Working against the water pressure, he finally got the boards out and water began to flow out into a large ditch and on through cuts in the ditch sides into the field we were standing in. By the time we got back to the truck it was ankle-deep and rising.

We had several rolled-up canvas dams in the back of the truck, so-called because they used to be made of canvas; now they were made of super-strong synthetic waterproof material, blaze orange in color. The fabric came in rolls, like newsprint, about five feet tall. It was hemmed with a loop on the top, to stick a dam pole through, and could be cut to any width depending on the width of the ditch it was to be used on, usually anywhere from six to eight feet.

Before we'd left we'd made up the dams with Keith's help, sticking dam poles from a stack near the barn door through the loops, then rolling another long pole and three shorter sticks in each one. I had no idea how any of this was to work.

Now Keith said, "David, grab one of those dams and follow me. You come too, Jerry. Bring those shovels."

I took a rolled-up dam and put it over my shoulder and followed Keith. The dams and poles were quite heavy; the larger ones weighed probably fifty pounds.

"Put that dam down and unroll it. Grab the brace pole—that's the other long pole, not the one in the dam. Put it across the ditch right here. You always want your brace pole at a point where the banks are sloping the right way, toward the downhill side. Okay, now give me the sticks." He was straddling the ditch, behind the brace pole. He dug the pointed ends of the sticks into the sediment on the upstream side at the bottom of the ditch and leaned the top ends against the brace pole. Then he came over, picked up the dam and got into the water upstream of the pole and sticks. He unfurled the dam, put the pole against the three sticks, and straightened out the fabric. He put one edge under each foot above the ditch, then took his shovel, put it against the center of the dam fabric, and carefully lowered it into the water.

The center of the dam submerged quickly and the fabric bellied out against the sticks. "Grab that rock there, quick, and put it in the middle of the dam, where my shovel point is. Okay, and another. Now put one on that corner of the dam, there. And another over on that corner." With that he moved his feet and his shovel. The center of the dam was holding and water backed up the ditch behind it. Keith watched it for a moment, then said, "That's got it. Now when you take a dam out to move it, you pick up one corner of it and lift until the water goes past it, then take out the sticks, roll it all up, and walk it down to the next place you want to set it."

There is something inexorable about water, particularly in the mountains in spring, and I was deeply suspicious of my ability to bend it to our purpose. It is stronger than any solid thing, flowing like fate over whatever lies downstream, finding the tiny recesses and depressions of our fear. It will kill you if you let it. But for an alert lifeguard, it would have killed me when I was eight, in the surf off the Gulf Coast. I'd grabbed a buoy rope and it came unhooked and swung me out into the deep water and under. Nebraska boys aren't used to that.

Since then, sweeping over slick stones in several rivers, it has tried a dozen times that I can think of. If it should ever succeed, perhaps here, while I am arguing with it, struggling with a headgate, trying to make it go in a new way, I wonder if I will see it coming.

. . . The dead sink
from dead weight. The Mission Range
turns this water black late afternoons

Richard Hugo wrote in "The Lady in Kicking Horse Reservoir." "No one cared the coward disappeared," he also wrote. That could be me, in the dark late-afternoon water of the Belts. They say a horse can sense fear. Perhaps water can too. I fear it. I admit it. I am a grown man, six feet three and two hundred pounds, and I fear quick water.

Jerry certainly didn't see it coming. We were setting a dam where Birch Creek runs under the road and into the field just west of the corner stackyard, and he took a step into what he thought was a shallow area and with a surprised look sank up to his chest. It was not a warm day, and the wind was blowing. Fortunately, it was lunch-time and we fished him out and retired to the bunkhouse, where I lent him a dry pair of Wranglers to wear the rest of the day.

That afternoon was spent memorably, with Keith. Bill had told him to get water flowing to the wild hayfields at the Stevens, and he took me along to help. Unlike the Birch Creek fields, which Keith knew perfectly, the Stevens presented a challenge. Since Bill had just leased this ground, it was as new to Keith as it was to me.

As we drove around the hayfields of the Stevens, I could not help noticing, again, that the ranch had not been run like Birch Creek. Strings from hay bales were everywhere. If Bill found one string on the ground at Birch Creek, he would have somebody's ass. "If any of you ever saw what one of these can do to a calf, you'd never drop one," he had told the crew over and over this winter. Old bits of machinery lay rusting in the field. And the place was littered with the carcasses of dead cows and calves. I'd never seen so many dead cattle other than at the gutpile.

"What did they die of?" I asked.

"Whatever," Keith said. "Cold, most of them. Probably some pneumonia in the calves. Some of these cows probably died of old age." We came to one creek crossing and Keith elbowed me. "Look, he's getting a drink," he said and pointed. A calf long dead lay with his bloated head bobbing in the burbling little creek, a macabre counterpoint to the mountain iris and daisies blooming on the banks next to him.

We walked the ditches, figuring out the way they were supposed to work. *Cherchez le* water. We crawled up ditch banks, through brush, crisscrossing the creek. "There's got to be a main gate system somewhere up here," Keith said after we'd crawled through a fence and along a larger ditch.

Finally we came to it, a relatively modern and quite complex confluence of gates and ditches and creek. Keith opened a couple of gates. "I'll go down this ditch, you go down that one. The water will be coming. Make sure it's flowing okay. Clean out any junk in the ditch that's stopping the water. I'll meet you back at the truck and we'll set some dams."

As I walked along the ditch, watching the water creep forward over last year's dead grass and leaves, I felt a few raindrops and looked up. One little cloud, still a lot of sun, and right in front of me, perfectly symmetrical over the ditch, was a startlingly vivid double rainbow. Just then I heard Keith shout, "Hey, dude, how's the irrigating?" I could see him, fifty yards away, walking along his ditch with his shovel across his shoulder, as I was along mine, soldiers in the same army. I pointed up ahead with my shovel, and he just smiled. Did he see the rainbow, too, or was it a personal hallucination? No matter. It was better than looking at the drinking calf. Anyway, as the ditch filled, I was too busy to look at either one. I was racing the water, mucking out debris, trying to stay ahead of the flow. I came to a huge jam, perhaps part of an old beaver dam, and I had to climb into the water and yank out branches and gobs of leaves and slime and god knows what until the water flowed correctly.

Then I met Keith back at the truck and we cut some dams from a roll of new fabric and put them on poles we'd brought and rolled them up and carried them to where they were needed. Keith was an irrigating wizard. He looked more like a mad scientist, hopping from here to there, fiddling with headgates, placing dams, making cuts in ditch banks with his shovel, just so. Watching him, helping wherever I could, I felt a little like I had that first morning, watching him catch a calf and doctor it. He expects me to do this?

Actually, yes. He left the last two dams for me to do, helping me as needed. I got a little overconfident on the last one. "Don't take your foot off there yet!" Too late. I lost one corner of the dam and it washed out. At least I know now what happens if you fuck one up, I thought, racing downstream to recover the sticks. I gathered all the parts and reassembled it, a good deal more carefully and successfully this time.

"I've got to get the backhoe over here, do some ditch work, move some of these cattle," Keith said as we drove across the fields. Over the past week Willie had been using the old International Harvester tractor on the Stevens fields, pulling a drag, a flat harrow made up of sections of chain. Its purpose was to break up the cowshit and spread it evenly across the surface of the field. The tractor was parked at the edge of a field that would be flooded in little more than an hour. "I've got to move that tractor anyway," Keith said. "I might as well drag it before I move it out of there." He parked the truck next to a large quadrangle of cottonwoods, obviously planted as a windbreak, but for what? Nothing remained inside them. "Must have been an old homestead here," Keith said. "Take that steel post there out if you can and open up the fence so I can get the tractor in here when I'm done." With that he walked over to the tractor and got it fired up.

I did as I'd been told, making room to get the big tractor across the fenceline onto higher ground. As Keith began his first turn around the field, I walked into the center of the square defined by the cottonwoods. I startled a red-tailed hawk, flushing him from the tree nearest me as I walked through, and I watched him climb in a wheeling spiral to the north. It was shady, maybe ten degrees cooler inside the tree line. These beautiful old trees were solid yet against sun and wind, doing precisely what they were there for, despite the fact that nothing remained of the house they sheltered but a few old bricks from a chimney and the traces of a foundation. Guards at the Tomb of the Unknown Soldier, screws at Alcatraz turned tour guides, doomed to protect nothing but an old idea.

I rode on the tractor for a few rounds with Keith. It was interesting at first—the roar and bounce of the tractor, the turds crumbling like stale crackers behind us, the cloudscape changing in the wind. But I knew, too, why Willie John derisively termed any chore involving driving a tractor "farming." It was so much less dynamic, more repetitive, and more *unnatural,* somehow, than most of ranch work. I know there is beauty in it; after all, I'm from farming country, in Western Nebraska, and I got my taste of combining as a kid. But I'd rather be chasing cattle ahorseback in a bitter wind than sitting and staring straight ahead in the numbing drone of a climate-controlled tractor cab.

For several days, whenever we had the time, Jerry and I had been making dams. That is, unfurling all of last year's dams from the place they were stored in the barn; checking the fabric for tears, and replacing it with new material if necessary; matching dams with brace poles and sticks, then rolling them up together like huge orange burritos and stacking them in the front of the barn.

Then one May morning Keith told Jerry to keep making dams

while Tyson and I started to distribute them. We drove around the ranch in the old maroon pickup to all the alfalfa fields. Since Tyson was driving, most of the time I was the one humping dams out of the back of the truck and laying them beside the head of each ditch, ready for the irrigators as soon as Keith got the water going. It wouldn't be long; for the past week, he'd been clearing ditches with a tractor and a device called a ditcher, which scraped the ditches clean of debris and loose dirt. First we went up to the huge, rolling hayfields on the Gurwell; then to the fields next to Bill's house; then to the field near the corner stack; then up on the bar—first to a strip of newly seeded alfalfa above the shop, then all the way along the bar on the south side of the road. This land had a complicated maze of ditches. "The water goes every which way in here," Tyson said. "Keith and Bill are the only ones who know it perfectly, and maybe Gary Welch, I guess." The fields stretched all the way down to the wild hayfields at Little Birch, quite near Section 19, where Willie and I had thrashed through the swamp.

Then, well beyond the antelope sign on the north side of the road, up above the Big Field, were a dozen more ditches going through a large undulating field called simply "the north side of the road." It was the most remote of all the fields up here on the bar. "I irrigated these a lot last year," Tyson said. "They go quickly. These ditches are pretty flat and your dam sets are pretty far apart."

I knew from what Keith had told me that the length of ditch between sets was dictated by how much land could be flooded evenly from each set, without overwatering or missing spots. The dams would need to be reset every twelve hours or so, early in the morning and at the end of the day.

It took us three trips with the truck to get all the dams distributed. It was enjoyable work, and it helped me understand the ranch geography even better, particularly in the areas where we hadn't fed hay. This place continued to amaze me in both its scope and its diversity, from the mountain country of the O'Conner down to the bottomlands; huge pastures of virgin grass like the Big Field and alfalfa fields like the ones on the bar; and the Gurwell, one of my very favorite places, with its wonderful hayfields and incredible views on the top, and its bluffs and coulees descending to the Smith River on the east, Birch Creek on the west.

At seven the next evening, Keith said, "You and Jerry go out on the north side of the road, remember where I showed you?" I nodded. "I set a dam at the end of the first ditch this morning. You need to move it to its next set. Then go over on the other side of the road, where those ditches come together, and move those two dams there." We zipped out on four-wheelers to the alfalfa fields on the north side of the road, next to the Big Field. It had rained a little more, and the road was very muddy, and so were we by the time we got there.

We took our shovels and walked out into the flooded land between the first two ditches. The water had spread out evenly for nearly a hundred yards. Walking down by the second ditch, when I found where the ground began to be dry, I walked straight back up toward the first ditch and planted my shovel in the ditch bank, the way Keith had showed me. "Here's where we want to move the dam," I said. The first ditch was wide and relatively deep, and it took a big dam and held quite a bit of water. So there was considerable pressure on the dam. When I picked up one corner, Jerry picked up the other, and the dam pole snapped in two, nice as you please.

Too late, I remembered what Keith had said about moving dams. Pick up only one side or the water pressure will break the pole. "Sorry about that. I'll go get another pole," Jerry said cheerfully, and hopped on his four-wheeler. "Thanks!" I said. "Get the biggest one you can find—it's a big ditch."

Twenty minutes later, it was nearly dark and I was worried about finishing the other dams we had to change. Finally I heard the whine of Jerry's four-wheeler. He cut off the county road and through the gate, holding the pole crosswise in his lap so he looked like one of the Flying Wallendas. "Kind of tricky, carrying this," he said.

"Nice job," I replied. "It should work fine." I'd removed the pieces of the old pole from the dam, and we quickly threaded in the new pole, set the dam and headed for the other field. Jerry had also brought a flashlight, and we parked one of the four-wheelers across the road with its headlights shining on the ditches we were working on. Something was wrong with one of the sets, I thought. There wasn't much of a place to go with it. Finally we got them done. Right, I hoped. Wrong.

I went out at six A.M. the next morning to get my dams done, mindful that it would take me longer than it should, so I'd better get an early start. The dam on the big ditch on the north side of the road was fine. I reset it—another long set. The ditch was really flat, which made it very efficient. There was enough water in the second ditch to start it going, so I picked up the dam I'd tossed out when I went around with Tyson, and got it set.

Feeling pretty good about things, I went to check the dams on the other side of the road, where I found Keith, fixing my mistake. "Remember I told you about this ditch, how it feeds the other ditch, and so you move the dam upstream instead of downstream? It's probably the only one you'll ever set that goes in that direction. That's why you had such a struggle with it last night. There was no place further upstream to put it." He shook his head. "No harm done, but you have to remember stuff like that."

The great thing about irrigating is also the bad thing: It lengthens your days, because you're out there early in the morning and late in the evening. Keith had told me when we started, "If you can get out there at daybreak and get it out of the way, you'll often get to change them again late in the afternoon and be done." Usually, though, other work kept me busy until evening anyway, which made for fourteen-hour days. And like feeding, changing the water was a constant that couldn't be neglected. We all had our dams to change, and if any member of the crew was absent, somebody else had to do that person's water, morning and evening, in addition to his own.

But the June mornings and evenings were so beautiful that I didn't much mind. Irrigating is a solitary pursuit, walking and studying the ground, and it was the best chance yet to encounter wildlife on the ranch. One night, above the shop, I'd almost stepped on a whitetail fawn, curled up in the long grass by a ditch bank. It bounded away, white spots on its rump gleaming in the low light. Another morning, a family of skunks waddled in front of me as I walked the same ditch bank. They most certainly had the right of way, and I yielded as courteously as I could.

Curlews seemed to love the irrigated lands, and huge flights of them would lift from the grass, their sharp cries filling the sky. Red-tailed hawks and bald eagles were common but still thrilling sights in the fields.

Every so often, you get reminded, in Montana, never to take the weather for granted. One morning in late June I stumbled out of the bunkhouse to irrigate at six A.M. in a T-shirt and denim jacket, as usual, and nearly froze.

It was cloudy and cold, maybe 45 degrees, and the wind was blowing to 40 mph. I didn't realize how cold it was until I was headed up the hill to the bar on the four-wheeler. The wind was like the ghost of March, cutting through my jacket. It was warmer once I got off the four-wheeler—until I got wet in the second ditch. I was on my second trip through the dams on the north side and what was by now second nature was suddenly almost impossible. Each dam took two or three tries before I could get it to hold; the wind would blow them out before I could get enough rocks or turf on top of the edges. That morning made the rest of the days changing water seem easy.

The irrigating days stretched into July. Everybody on the crew was tired of the long hours, pulling on boots still wet from the morning and walking ditch after ditch, land after land. Like everything else I'd tackled on the ranch, though, there was a feeling of accomplishment in it. I was getting better at setting the dams, and faster—which was fortunate, because most days I'd have nine or ten to set. There was a knack to walking the fields, finding the water's path and figuring where it needed to go next; to digging cuts in the ditch banks at the right intervals; to making the final sets at the wide mouths of the ditches, moving the water onto the next land and the next and the next.

I knew I'd earned something special with the work, this being here now: the smell of alfalfa, the rainbow arching from Kite Butte up to somewhere high on the O'Conner, the chunks of rust-and-green-lichen-covered shale and granite with their darkened parts embedded deep in the black earth, more under the water, rough gems at the bottom of the ditches.

One night a covey of Huns whirred up from a ditch bank in front of me, eight fat, rather ungainly little partridges indignant at being rousted, gray bundles limned with russet. Partridge, *Perdix,* from the Greek *perdesthai,* to break wind, I was reminded as their distinctive whirring filled the evening stillness. I walked on with my shovel across my shoulder, looking around in awe, fatigue dispelled, my legs suddenly light as the alfalfa whispering against them in the breeze.

I have a farmer's pride at seeing the irrigated alfalfa grow tall. I flew again one afternoon with Bill, and the water glistening below the green leaves on the flooded sections was a satisfying sight, particularly when he said it looked like my dam sets were okay. "You can tell how it's doing by the color," he said. "If it isn't getting enough water, it's really dark green. If it's getting too much, it's pale. If it's just right, it looks like that."

By then we were getting ready for haying, which meant a lot of shop work. We serviced the Dresser loader and did extensive work on the old Chevy loader, a custom-built rig with a Chevy 454 engine, an open cockpit and a big set of forks on the front. It hadn't been run much since last summer, and it was in tough shape. Keith did some welding on the frame and welded the gas pedal back on while Willie and I overhauled the brakes, changed the hydraulic fluid, put in a new battery, changed the oil and filters, and greased it.

We serviced the Ford tractor, greased the balers and loaded each of them with sixteen new balls of baling twine. Keith tuned up another homemade open-air vehicle called a jitney, which looked like a cross between a Checker cab and an old Land Rover. It had a set of forks on the front and could pick up only one bale at a time, compared with two for the loaders. It is a favorite of Doran's; he has run it for the last few years. He worked on repairing a slow leak on one of its tires and helped his dad change the oil. We serviced both swathers and changed out the blades on one of them, and we went over the retrievers carefully, replacing all the fluids, looking for cracks on the frames, changing out the rear tires on the diesel.

Now all we needed was the hay, and with the combination of sun and moisture, it would be time to cut it soon.

Many of the fields were irrigated again after cutting in an effort to get a second growth worth haying. Those mornings, irrigating in the coolness and solitude of daybreak, were some of the best time I've spent on the ranch. The moose often came out of the creek bottoms to feed on the alfalfa stems on the north side of the road, where I irrigated. Usually, when I headed up the bar road, either on a four-wheeler or in the maroon truck, I would see them in the fifth, sixth and seventh lands. With their heads down they looked like dark brown boulders in the distance, betrayed only by their location in the unnaturally manicured expanse of green. Then, as I got closer, I could see their great hocks planted like saplings in the earth, and the swaying of their withers as they cropped and chewed. I counted as many as eighteen head. One morning a cow and a calf that were close to the road spooked and jumped the fence in front of me, wild-eyed and indignant, but most of the time none of them were bothered by my puny presence as I passed them and moved my dams on the other side of the field. Invariably, by the time I was finished setting dams they had gone, melted soundlessly into the thick brush down by Little Birch, I suspect.

One of the last irrigating mornings, as I drove up the bar road at daybreak, I was surprised not to see one moose. It was the first time in a month they hadn't been in the field as I passed. But when I walked over the hill to the business end of the second land, the morning turned weird in a hurry. The moose were there, just over the rise where I couldn't see them from the road, and I walked practically on top of a cow and a calf.

I backed up slowly until the cow lowered her head and snorted, at which time I abandoned all pretense of composure and ran like hell for the pickup. I looked back once and wished I hadn't. She was gaining, and she looked very angry. She was snorting, still, and a huge sling of something viscous—snot? saliva?—was trailing from her nose and mouth, glistening in the sun. Thankfully, she was satisfied with ejecting me from the premises, and when I got in the truck, she cantered around and went back to her calf. I decided to enjoy a few moments of the morning from inside the maroon pickup and congratulated myself on setting the amateur record for the 100 meters through alfalfa in hip boots, carrying a shovel. Pretty soon, maybe I could even breathe again.

JUNE 15

Tonight I changed my water in a hailstorm. I saw it coming from the west as I made my way between my fourth and fifth dam, out in the middle of absolutely nothing taller than the calf-length alfalfa. So I just set my dam and kept on going, hailstones tattooing off my head and shoulders, purple alfalfa flowers nodding like addicts under the barrage. It passed quickly, leaving the field white as winter, but undamaged as far as I could see. Now the storm became a spectator sport. I watched the cell move past to the southeast, over the Big Field, seemingly gathering intensity as it went. Lightning jabbed at the hills from the big charcoal cloud, and the thunderclaps all but drowned out the cries of the curlews that wheeled overhead. Just another evening in big country.

chapter 15 • more water, please

BILL WAS MOVING AGGRESSIVELY TO MAXIMIZE HIS USE OF THE leased ground, and that included water projects on a scale I wouldn't have thought possible. By far the most ambitious was a project to improve a sizable dam and its reservoir, to make it safer and more useful for irrigation. When I heard what Bill had in mind, I had a hard time believing it. Among other things, he proposed to dig a hole through the face of a dam that happened to be holding back a lot of water.

As he discussed this with me, we were standing on top of the dam in question, which held Keep Cool Reservoir. Mountains showed fresh snow on the horizon; nearer to the water, a single wind-bent juniper grew out of the side of a large hill just as green and no doubt just as overgrazed as the rest of the Manger lease.

"How old is this dam?" I asked. "I don't know, maybe forty-some years," Bill said. "But the spillway's damaged, and it really needs to be redesigned. So we're going to raise the dam, build a new spillway over here"—he pointed in a large arc—"and then we're going to punch a hole through the dam and put a big culvert in, an intake over here, a riser with a shutoff valve right here, and an overflow down the back side, to there."

"Who's we?"

"Me and the crew, which means you," he said.

"Hey, I didn't sign up for the Corps of Engineers," I said. "I thought this was a ranching job."

"All part of it," he said. "There's nothing more important to the ranch than water."

"Back to this culvert," I said. "Will it go above the waterline?"

He chuckled. "No, below. About eight feet below. Grab that propane tank, will you? Let's burn some brush."

This guy's crazy, I thought. Below the waterline? How do you do that? I was distracted from thinking about it, though, because Bill Galt said, "I take the burner, you take the tank. The hose is only four feet long, so you have to follow me closely. Let's go."

I definitely got the raw end of this deal. The tank was full and it weighed maybe seventy-five pounds. Plus Bill was moving very quickly through the brush on the hillside behind the dam, shooting flame out of the nozzle of the propane torch. Blasting away with what amounted to a flamethrower, he was irritated because the spring moisture had dampened the brush enough that the fire was not spreading well. "I guess we just can't get it going," he said as I walked through flaming brush he had just torched.

I was quite thankful when that experiment ended. It allowed me to kick some dirt on my smoldering boots and begin worrying about this dam situation.

"Here's Otto," Bill said as a pickup truck pulled up near the dam. "Give him a hand, will you? He's got to flag some contour lines for the cut we need to make for the spillway."

Otto Ohlsen was an engineer, but he didn't look like a desk jockey. He had the deep tan and lithe build of someone who spends his life out of doors. And there was a precision about him that complemented the weathering. A faded Patagonia shirt, well-used hiking boots and a measured stride. Discerning eyes behind round wire-rim glasses. This man looked at these hillsides and saw water and earth, angles and gradients, weather and time.

He worked for the federal government—the Soil Conservation Service. As a conservation and safety project, this dam job qualified for federal assistance, including Otto's expertise. Otto has worked with Bill on other projects, and since Bill is also on the local SCS board, he knows him well.

"We have a lot of old dams like this in Meagher County," Otto said. "Many of them were built by the old Civilian Conservation Corps crews in the Depression. Many of them are classified as high-hazard now. It's quite a concern."

First he needed to survey the area behind the dam where the new spillway would go. He picked a reference point, high on a knob above the existing spillway, and set a revolving beacon on it that projected a light beam. Then he handed me a long, telescoping pole with a photosensitive receiver on it. He would sight with his surveyor's glass, then tell me where to stand and I would raise or lower the pole until it intercepted the beam of light. Then he would note the height and we'd plant a little wire stake with a red plastic flag on it at the spot where the bottom of the pole was. We repeated this procedure down one hillside and back up the next, then moved over a few yards and did it again, so we could mark the depth of the cut Bill needed to make with the D-8 to contour the spillway correctly.

As we worked I thought about how many times this scene had been played out across the West over the past century and a half, men trying to take the measure of the country, to reshape it, change it, gentle it for their own purposes. For shelter, for water, for transportation, for wood, for gold and silver, for the satisfaction of dreams that were often dreamt back East by others. I was not deluded into thinking that Otto Ohlsen and Bill Galt were not doing the right thing here, making this lake useful instead of dangerous, saving water that had been treasured for the past century, named by shepherds who would put their milk and other provender in the bubbling spring to keep it cool. But I was glad there were a hundred thousand acres or so under Bill Galt's control that would not be seen through a surveyor's lens in my lifetime.

At the same time, I couldn't help but be impressed with Bill's aggressiveness and determination. This ground had been leased a long time, ranched a long time, by others. He had leased it at a somewhat more favorable price than his predecessor, and he could simply have put what cattle he could on it and hayed what he could of it.

But not a week after he gained control, he was already embarked upon an ambitious plan to make of this more than it had been. This was not a small project. Tons of earth had to be moved, and Bill was frankly gambling that he could get the culvert into the dam without major mishap. If the face of the dam was to give way, he would have a roaring flood instead of a lake on his hands. I asked Otto what he thought, and he chuckled. "Bill likes to move quickly," he said. "This should be interesting, and unless we get a big rainstorm at the wrong time, it should work."

We certainly weren't going to wait and see what the weather had in store. Bill started excavating immediately with the D-8. Keith worked with the D-3 and a backhoe. "You want to see something special, watch Keith on the backhoe," Tyson told me. "He can do absolutely anything with that thing."

Indeed, Keith was nothing short of a virtuoso. The Case Extend-a-Hoe had a bucket on one side of the cab and the hoe, a big scoop with teeth that chewed out dirt, on the other. It ran on wheels rather than on a track, but was immensely maneuverable because of two booms that swung out from the sides, big flat plates that the operator could plant for stability. Keith could also move out of tough spots by planting the hoe in the ground, lifting the body of the hoe up and leapfrogging forward on the booms, as though he were using crutches.

I could have watched him for hours, except I had my own piece of equipment to run: a shovel. As Keith began to cut the trenches for the culverts, a lot of dirt and rock had to be moved by hand, or rather, by back. I had little time to think about anything but digging. Another romantic day in the cowboy life. Yippee-ki-yay, dig those ditches.

Before long, the trenches were ready for the riser and for the first piece of the big drain culvert on the back side of the dam, so Keith got the chore of bringing the fifteen-foot-long section of culvert down the steep, twisting track from the pasture near Nancy Manger's house to the dam. Backwards. The switchbacks were so sharp, and the culvert so long, that there was no way to drive down in forward gear. I got in with Keith to help him navigate, which turned out to be like helping Michael Jordan make a layup. He made it look easy, sailing down the mountainside backwards with the enormous culvert sticking out the back of the truck, picking a line into each corner with an eye to getting lined up for the next one. As we approached the dam, he said, "Maybe I should gun it and give 'em a thrill."

"That's okay," I said nervously, looking in the side mirror at our right rear tire, maybe ten inches from the edge of the cliff. "A little close over here."

But by the time I'd said it, he'd already corrected to the center, and in another couple of minutes Jerry, Tyson, Willie and I were lifting the culvert out and putting it by the side of the trench. After a good hour's shovel work to finish the trench preparation, we rolled the culvert in, then worked on getting it fastened to the riser, which Keith had maneuvered into position with the backhoe. The culvert sections were joined with bands that would line up only if the joint was exactly straight, which meant wrestling with the culvert, digging out high spots and filling in low spots. Then when the alignment was right, the bands would bolt on correctly.

It was late in the afternoon when we finally got the culvert banded to the riser. Then we had to bed the culvert by hand, digging dirt from the sides and packing it around the bottom wherever it was needed. Then Keith bladed in a little dirt, and we tamped it down, first by hand, or rather by foot, and then with a torture device known as a Whacker Packer. This was a motor-driven tamping machine that vibrated like a jackhammer. I had definitely aggravated my already-sore back, wielding a shovel, and as I ran the tamping machine, each vibration was agonizing. I did my best to ignore the pain and kept Whacking away. This is known as cowboying up, and it is encouraged.

A few days later, as we got ready to head to the dam after feeding hay, Bill radioed Tyson and said, "Take twenty sticks of dynamite and some blasting caps with you. We've got to clear a couple of beaver dams." So Tyson, Jerry and I headed down the road in the old maroon truck, now outfitted much like a Hezbollah suicide bomber's vehicle. We had two full tanks of gas (one of which had a leak) plus two five-gallon jerry cans on the back, along with enough dynamite and detonators to blow us and whatever we ran into to Allah and back. As if that weren't enough, the steering and front end were still horribly loose, the brakes were still very spongy, and the big spray tank on the back was three-quarters full of water, making the truck lunge and lurch like a wounded bear.

We made it without martyring ourselves, and pretty soon Bill was on the D-8, Keith on the D-3, and Tyson was driving the big Cat scraper, with which he scraped up loads of dirt from a "borrow" on the hill above the dam and brought them back for the dozers to work into position. Willie, meanwhile, was using the grader to smooth a couple of other roads on the Manger lease, so that left only Jerry and me to do the tamping and shovel work.

After lunch Keith grabbed me to help him assemble the rest of the culvert for the overflow. The last two sections of culvert ran above ground, down a steep hill. The end of the second section flared out horizontally, just like a gutter drainpipe. The first piece of culvert wasn't too bad. We humped it from the truck (Willie John had gotten the job of backing it down the road today, and succeeded, with slightly whiter knuckles) to the top of the hill, then tipped it over the edge. The hill was very steep, so it was hard to hold in place, but when we got it lined up, I held it and Keith bolted the bands on.

The last piece was something else. It was slightly too long for the remaining slope, so Keith had to dig out the pond at the bottom of the hill with the backhoe. Then he swung the hoe over to me, where I stood thigh-deep in cold muddy water, and I chained the end of the culvert to the hoe. Then Keith tried to wiggle the culvert into position. The first few times we tried it, the culvert refused to co-operate, and I had to unchain it so Keith could dig and scrape underwater. When we finally got it close, Tyson came over and helped me wrestle the culvert into alignment, and after a couple of hours we managed to get bands around it. The culvert had a noticeable bend in it when we were done, which I correctly suspected neither Bill nor Otto would like very much, but it was done, and it would work.

At sundown Willie John and I gassed up the D-8 and the scraper with Keith's service truck. It would have made a great commercial for Caterpillar, the trademark yellow glowing to gold in the low light, contrasting against hills, lake and lavender sky.

"Willie, you guys get that dynamite and meet me at the creek," Bill Galt radioed, and soon I was standing by the largest beaver dam I'd ever seen. It had backed the creek up into a huge, eddying pool. I thought it quite ironic that we were setting about to destroy a perfectly built natural dam even as we were repairing a flawed man-made model, but in order to get the irrigation system flowing correctly here on the Stevens, this great wall of sticks and brush had to go. The dam builder was nowhere to be seen as Bill Galt nestled four sticks of dynamite into the mud beneath the main part of the dam and placed the fuse and detonator, and strolled onto the nearby hillside with the rest of us. A muffled boom and a shower of sticks, mud and water perhaps sixty feet high pleased the spectators greatly, but upon inspection the structure of the dam was found to be essentially intact. Another four sticks, placed closer together, did the trick in even more spectacular fashion. Within ten minutes the dam pool had drained and the creek seemed to be running normally. "He'll probably have it rebuilt in a couple of days," Bill said cheerfully. "They usually do."

We got a nasty rain, just as Otto had feared, which made the lake rise precipitously and turned the whole project into a quagmire. But after a couple of days spent in the shop, that damned dam, as the crew referred to it, again occupied most of our time. The culvert running through the face of the dam was shorter than the outflow culvert, but a whole lot trickier to install. Keith outdid himself, scooping out untold yards of dirt and making himself a platform called a cofferdam stretching out into the lake, from which he could work. Give him a spot to park his backhoe and he could dig up the world.

Finally the trench was long enough to stick the culvert in. Working about ten feet deep in the trench and at least eight feet below the waterline, Willie John and I rather hurriedly bolted the last three bands on, watching the mound of earth standing between us and a killer deluge. As Keith scraped away the dam face with his backhoe, the color of the clay at the bottom changed from reddish-brown to dark gray, and the surface seeped and trickled, making all of us, particularly Bill, exceedingly nervous. We got the culvert bolted in, got it bedded, and got the hell out of the trench. Keith and Bill used the dozers to cover it up with dirt. We tamped it and they covered it again.

The excavation site was sort of like the *National Enquirer,* with a seemingly endless appetite for more dirt. The more we tamped and the more loads Tyson brought in the scraper, the more it seemed to need. I spent one afternoon removing huge rocks, by hand, from the spillway area where the scraper was operating, because they got in the way of the huge, lumbering machine, and it was very difficult for Tyson to avoid them. Picking rocks and lugging them out of the scraper's path, I found, was disgustingly good exercise.

Finally Keith scooped the dirt out of the area in front of the intake end of the culvert, and water began to flow into it. "Open that shutoff valve," Bill said. I turned the big red metal wheel at the top of the riser, and water began to shoot through the overflow culvert into the little pond behind the dam. The system we had installed actually worked.

Ambitious water projects have often not treated Montana ranchers kindly. Ever since the Desert Land Act was passed by Congress in the 1870s, people have been trying to bring order to the chaos that is Montana's water supply. The act provided offered a section of land for twenty-five cents an acre to anyone who would agree to irrigate it within three years—similar to the Homestead Act, which had a five-year residency requirement. Much of the time, three years passed, no irrigation took place, and the government got back an unwanted, overgrazed square mile of ground.

The west and central parts of the state are particularly well endowed with water—after all, both Missouri and Columbia rise in the region—but harnessing it for irrigation has never been easy because of the wild, unpredictable nature of many mountain streams and because of the weather, which frequently brings on the extremes of flood and drought, but less often provides a workable happy medium. As water demand has increased in recent years, so have the horror stories. The Jefferson River, near the point where it joins with the Madison and Gallatin rivers to form the Missouri, has been badly dewatered to feed big irrigation systems. The Smith and Shields rivers in Meagher and neighboring Park counties also suffer badly in drier-than-normal years.

So Bill was doubly blessed to have plentiful water most years and systems in place to handle it, but the Manger was something else again. In addition to the reservoir, the large pivot and wheel-line irrigating systems needed a lot of work, and then there was the Dry Range.

Bill collected me out of the shop one morning soon after the dam work was completed. "If you want to see some country, you'd better come with me, but you're going to have to do some work."

"I hate the sound of that."

We drove toward the Stevens, but instead of turning in, we went a little farther toward Lingshire, turning toward the mountains at the Clear Range turnoff. We drove up through Doggetts' land on the Clear Range, up toward the trees that mark the start of the Dry Range. Soon we were in switchbacks. The air cooled a bit as we gained altitude, and paintbrush and lupine blazed in profusion along the road.

We came to a cattle guard with a broken gatepost next to it. "Make a note of that," Bill said. "You'll come up here and fence this." A few yards farther, Bill turned to the right on a tiny track, and parked next to a rusty green storage tank. "This is Den Gulch Spring," Bill said. "The spring was first improved sometime around the turn of the century. We need to get it working better."

These mountains are not called the Dry Range for nothing. Because of the lack of water, few cattle have ever been on this land. Henry Lingshire and the Walker Land and Livestock Company ran sheep. So did the Mangers, and so did Bill Loney, when he had the Manger lease before Jim Witt. Rancher Art Watson, who wrote several excellent books about early days in the area, including *Devil Man with a Gun* and *Montana Trials and Tribulations,* ran horses on the Dry Range. Horses require less water, and will travel miles to get it; in the winter, they can get their moisture from eating snow.

Still, I thought about Bill's improvements at Keep Cool reservoir, and figured Bill had something in mind to defy history and get water to his cows. The area around the tank was covered with old lumber, hoses, rusted-out troughs and other junk. I began to pick all the garbage up and started a junk pile.

Bill checked the water level in the tank—it was about half full—and he and I walked fifty yards uphill from the tank to the spring. Water bubbled out of black mud, forming a small pool. A crude siphon and pipe system ran up the hill and into the tank. Next to the tank sat an old pump and an engine to drive it. A high-pressure line ran into the ground from the pump. Nearby was a small stock tank once hooked directly to the spring with galvanized pipe, now rusted and interrupted, ends sticking crazily in the air.

"The siphon brings water into this tank, and the pump takes it somewhere over there"—Bill waved toward the mountain behind us—"to a bigger holding tank, higher up."

Bill seemed to be pleased with what he saw. "I'm going to get a new motor up here. I hope we can salvage the pump. We'll dig out the spring a bit with the backhoe and put a better siphon in there. Then we have to trace the high-pressure line and see what's happening with it."

Over the next few days I helped Keith a little with the pivot, fixing crossings and culverts, but worked much more on the wheel line with Donnie Pettit, who would run it after we had it ready to go. Keith and Donnie and I had located the pipe for the wheel system's main line in a jumbled heap near the barn last week. We'd sorted it and brought a few sections in need of repair back to Birch Creek, where Keith and Bill had spot-welded them with a wire-feed welder that worked on aluminum.

This morning I went into White Sulphur with Bill, and he bought two stout new Briggs & Stratton pump motors—one for Den Gulch and one for the wheel line. We stopped in at Edward's Grocery so Bill could buy lunch for the crew. I'd never been in Edward's before, for the simple reason that it was always closed long before I got off work. It was a wonderful, old-fashioned grocery store, I found, with a terrific meat counter and good produce. Bill got a few packages of salami, a head of lettuce, a loaf of Wonder bread, mayonnaise, Pringle's potato chips, apples and oranges, and a package of plain cake doughnuts from Eddy's Bakery in Great Falls. And, of course, a twelve-pack of Pepsi.

"Anything else?" he asked.

"Yes," I said, and grabbed a jar of mustard. "How you can eat salami sandwiches without this stuff is beyond me."

"You city types have strange ways," he said. June Bergan loaded the groceries into a box and smiled at us. Bill signed the slip and we were away. A gentler, simpler era preserved.

Once we got to the Stevens, we loaded all the pipe on a trailer and took it down to the field. We took one of the new motors out of the back of Bill's truck, and Bill, Donnie, Keith and I worked on getting the pump going. This involved priming with a hand-operated pump handle, which I did.

Then Donnie and I unloaded the twelve-foot sections of pipe, got it lined out in its approximate position and started coupling it. "You guys get this thing going," Bill said, "while Keith and I take this motor up to Den Gulch." He drove right through the field with the wheel line and out to the Clear Range road.

The main line was the easy part. The wheel line stretched about a quarter of a mile on each side, and each section of it needed to be hooked up and checked over. We started getting them into position, checking the nozzles and the fittings. A few of the wheel sections had blown off the ties that held them and rolled a couple of hundred yards downhill. Nothing to do but push them back by hand. I wondered how good this field could be. At the moment, gophers and mosquitoes were the only things prospering.

Getting the sections lined up perfectly for coupling was the hard part. The sections were heavy and didn't move well laterally, and lifting them up even an inch or two was also difficult. Amazingly, we found all the coupling collars but one in the grass; Donnie said he knew where to find one in the shop.

Once again I'd been dry-gulched by somebody taking off with the pickup containing my water. This time it was Bill, who also took lunch. Watch that in the future. Donnie and I finished his large thermos at noon and swore the rest of the cloudless afternoon. I briefly considered risking giardia by drinking from the creek, then remembered all the dead cows that had wintered right in it and stayed parched instead.

Donnie was great to work with. He was a lot like Keith Deal—handy at everything—but, not burdened with the foreman's responsibilities, he was a lot more relaxed. He told me a little of his background as we wrestled with the big wheel sections.

Raised in tiny Cascade, about fifty miles north of the ranch, between Helena and Great Falls, he worked on the well-known Hodson ranch nine miles out of town during high school. When he got out of school, Hodson paid him three hundred dollars a month to work full-time.

"Doing what?"

"Branding, calving, roping and greasing cows, breaking teams of draft horses for feeding hay."

"Greasing?"

"Rubbing stuff on their tits when they were chapped. Yeah, those old boys at Hodson's didn't believe in taking it easy on high school kids. I'd be greasing a cow and they'd cut her loose and we'd have a rodeo. Or I'd rope a calf and they'd take my rope and cut it up and make lead ropes out of it.

"I was at the old Gaspar place at Hodson's one day in the winter of '78–'79 when the big snow came in, four foot of it overnight. It was cold, too, maybe thirty below, with the northeast wind blowing. I was in my bunk early that morning and I heard some of the old hands talking, and they were saying they didn't think we could get out to the cows, so I just stayed in bed.

"They hollered at me about three times to get out of bed. Then

Louie Eller, one of the hands, showed up next to my bunk with a shotgun. 'Get up,' he said, and when I didn't, he shot once over me. When I stayed in bed he shot me right in the ass with birdshot."

Donnie shook his head and grinned. "I've never slept in since." He thought for a moment, and added, "I got them old boys back, though. One night it was about ten below. They were sleeping in the lower bunkhouse and I shut the heat off on them. They didn't realize it until the next morning, and they like to froze."

He laughed. "That Louie Eller was a hell of a hand. He was the cow boss then, and one day we was moving a bunch of cows and Louie come roaring down off a hill and his horse stumbled and throwed him and he broke his leg.

"Well, we tried to pick him up and take him in, but he wouldn't have it. He said, 'Keep going with them cattle, and when you get 'em moved, come back and get me.' There was nothing we could do. So he laid there all day. It was evening before we got those cattle situated, and he was pretty bad when we went back and took him to the hospital. There wasn't anything he could say, though, because that's what he told us to do. He ended up losing the leg. He didn't blame us any. He's still around, too; we're good friends."

I couldn't think of a response to that. A few minutes later, I asked him, "How long did you stay at Hodson's?"

"Till I was about twenty. Eight or nine of us got canned from there one day, drunk. So I went to work for a horse trader up at Great Falls, Tom Pettipiece. I'd met him around Hodson's. I'd ride the horses for him, show them to people. That's how I met Billy. He came up there to look at horses and I showed him a pretty good bay gelding, name of Petey. Bill bought him.

"Then the next spring I was looking for something to do and I came out here to Birch Creek and asked Bill for a job, and he hired me."

"So you've been here ever since?" I asked.

"Shit, no. Billy and I get sick of each other every so often and I go down the road. I've probably quit him half a dozen times, but I always come back."

That made me think of Keith. "You ran the crew for him for a while?"

"Yeah, but that got old. And when I came back last time, I think it was '93, he asked me to go to Lingshire to calve out his cows." Donnie always had a smile, often accompanied by a ribald joke, but he never slowed down a step all day, and I stayed right with him and felt good about a day's work when we finished.

His four cowdogs—Red and Socks, the seniors, and adolescents Fred and Bear—had no compunction about drinking from the creek. Mostly, though, they watched us from underneath Donnie's pickup, the only shade they could find.

We serviced the two motors that propelled the wheel line—one on either side, about midway along—changed oil, replaced filters, and checked belts and hoses. Bill came back through around three. We'd left an opening in the wheel line so he could drive through with his truck instead of having to go all the way around. He was pleased with our progress.

While Donnie changed the water on the wild hay I made small headway cleaning the shop at the Stevens. It was going to take a lot more time than I had this afternoon. It had been years since the doors had been closed; blown-in dirt had been snowed upon, turning it into mud, which had dried on the floor. An amazing assortment of equipment, mostly junk, some of it useful, littered the place, scattered by time, frustration and carelessness over every available surface, including the floor. It was hard to see where to start, but I shoveled out the double doorway so it would close, cleaned off the top of the long workbench and began to try to organize parts. Irrigation systems in one corner; hydraulic cylinders in another. Electrical stuff on one side of the bench, truck and tractor parts on the other. I barely got a foothold, but it felt good to start.

A door led from the back of the shop into the barn, and I couldn't believe what I saw when I opened it. Broken boards, broken-down equipment, putrid cowhides, filthy pulling chains, rotting grain spilling from torn sacks. Rat droppings everywhere. There was another enormous project, and I knew it would soon be mine to tackle. There was no way Bill Galt would stand for anything like this on a spread he was running.

I rode back to Birch Creek late with Donnie. I had to hustle to get my water changed before dark—especially because I had to swap out a dam at a trouble spot on the bar.

At the bunkhouse, sunburned, mosquito-bitten and sore, I thought about what it took to get food and water to cows, and how puny my efforts were in relation to all that was needed. Everything I'd done today and most of what I'd done since I came to the ranch revolved around filling those simple needs. No wonder so many ranchers without the acreage or resources of Bill Galt went broke or gave up the struggle. Not to mention some who had both, and still lost it, like the Mangers. At one time the Manger spread had been one of the largest outfits in the county. Much of it had been sold off, and the family no longer wanted to run what was left, so they had leased it and let it run down. Now they were counting on Bill's enterprise and energy to make the remaining ground pay—pay him, and them. After what I'd seen of the place, I figured Bill was one of the few operators who had a chance to make it work. This year, I knew, he would probably lose money on the lease, despite the concessions he had negotiated, but as he got the place squared away, subsequent years should be better.

I flew with Bill again the next day, this time to get some parts for the pivot system. We made the flight from White Sulphur to Gallatin Field outside Bozeman in just under half an hour. It took just a few minutes on the ground to get the parts, and after Bill took a lustful look at a pristine new Bell Jet Ranger helicopter at Arlin's Flying Service, we completed the round trip. As we landed, a Meagher County sheriff's deputy met us and asked Bill if he would mind flying over Mount Edith, one of the snowcapped Three Sisters that topped the Big Belts. Both a funnel cloud and a lightning strike had been reported in the area overnight, and the department wanted to check for damage and/or smoke.

"Sure," Bill said, glad as ever of a reason to stay in his airplane. Keith, waiting for us at the airport, took the parts, and we took off again. "Your job is to tell me if I'm about to fly into the side of a mountain," Bill told me as we headed over the ranch and toward the Belts.

"Don't worry, you'll hear from me."

We climbed steadily, flying over heavy pine forest, then several appetizingly bright blue mountain lakes, then the bare rock and snow above the timberline. We made several passes and saw no hint of freshly downed trees or smoke.

"It's worth the ride to see this country up close, isn't it?" Bill asked.

"Absolutely." I pointed to a tiny lake high on the slope of Mount Edith. "That looks like a glacial cirque."

"Yep, I think it is," Bill said. "Wonderful country."

We flew west, and had a look at Newlan Creek Dam and reservoir and at the Stevens, then circled back to White Sulphur with a negative report for the sheriff.

Bill got his flying in just in time. The clouds returned in the afternoon, and it rained until nightfall. Somehow irrigating in the rain seemed redundant. Also wet.

Keith and I headed straight up to Den Gulch after changing water in the morning, with his pickup and the backhoe. He bladed off a flat area near the tank, and we placed two fifty-five-gallon drums of gasoline on it and hooked them up so the motor could run for long periods of time. We also built a rail fence around the big storage tank, the engine and the pump, to keep cattle away from the machinery. Donnie would be surprised, I thought, next time he came up here to check the level in the tank.

Then, feet sunk in black muck, water up to our thighs, we installed a riser and siphon in the middle of the pool at the spring.

"That ought to do it," I said.

"I don't know," Keith replied. "I hope so. This isn't going to make any more water in the ground. I wonder how long this spring will run strong enough to keep water in the tanks."

A PLACE TO PUT THINGS

Stalls where horses dead now fifty years
ate oats grown in forgotten fields
are stalls still, rough-notched and square-nailed.
Doors hang on hand-forged hasps and hinges,
harness yoke and bridle
with mended rein remaining.
Stalls still, filled with things left idle:
A flat-head six that knew too much
of heat and cold and country. Swather wheels
that turned on ground
that turned from range to barley. Driveshafts
welded shut to pound steel posts,
rat-chewed canvas dams,
fencing pliers rusted open, fifteen
shovel handles, busted. Hip roof stops only half
the sky, so this became a place to put things
halfway cared about. Sun and snow and birdshit
fall on floorboards ricked and gaunt
as an old cow's backbone. No love
is made in the loft. Nothing up there
unless you count dented irrigation pipe
and the rear end of an Allis-Chalmers
no person living on this outfit
ever drove. Still good but long ignored. New things
break quickly. Old things get older,
fall from favor, purpose leaked away
with someone's faded plan to turn work and land to cash,
buy a pretty dress for a redheaded daughter gone.
Ninety years in the lee of a little hill
in the Belts somehow made this old. Knapweed
pokes through knotholes in the north wall. Deer
still come in the evenings to find spilled grain.
The new tractor spends nights elsewhere. Horses
graze the hillside, and the stalls no longer hold them.

chapter 16 • thirty-six square miles

"Under endless skies
I'll ramble along
with miles and miles
of room to roam."
—WYLIE GUSTAFSON, *ROOM TO ROAM*

OF ALL THE RANCH, THE BIG FIELD WAS THE PLACE WHERE I could best get a sense of what the open range must have been. It was the place where this spring Bill Galt had said, "This is what cow country should look like." When he came to Birch Creek, it was fenced only around the perimeter. Inside that fence was an entire township—thirty-six sections of ground, each one six hundred and forty acres, a square mile. Twenty-three thousand forty acres, thirty-six square miles, without a fence. Since then, he had fenced off a large pasture called the Dry Gulch, but the cumulative effect was still profound. It was rough and immense, with grass, sagebrush, rock formations, mountains and considerable water. It was a good place to put cows and a tough place to gather them, and it was not a place that easily forgave fuckups. Hit a rock and crack a crankcase or fall off a horse and break a leg, and you would be there for a while.

The cattle were wilder in here; the open space made them much warier of people and much harder to control. Something about the immensity of the place affected me, too, perhaps because whatever you did here—fencing with Wylie in a blizzard, rebuilding the Rock Springs branding trap, gathering cows—was bound to have an increased element of challenge. This ground was not tamed.

The branding-trap project was educational. Keith and Willie John both seemed to have their noses a tad out of joint that Bill had taken me up there to explain the job, instead of one of them. They didn't like hearing from me about what needed to be done.

We loaded the little skidsteer loader with the auger fitted to the front in a trailer, with all the usual fencing stuff plus a few panels, a dozen beefy posts and two dozen long rails. Willie John met us out there. He and Keith discussed the job as if I weren't there. When I told them what Bill said about backing up a couple of the brace posts with larger posts rather than replacing them and having to rewrap all the fence wires, Keith snorted. "If we did that on our own he'd kick our ass," he said. He dug the postholes and headed back to Birch Creek.

It was sunny and hot. Willie John was quiet at first, but he seemed to warm with the day. He put Jerry to work repairing panels, which was a challenge because it's very difficult to drive nails into old weathered wood. We started replacing rails. They had to be notched with a chain saw, then pounded in with sixty-penny nails. Willie poked some fun at my technique, which was admittedly inefficient. I choked up on the hammer about halfway because I couldn't control the larger swing you got holding it out at the end.

He said, "Do it like this." He held it out at the end and took it back way past his shoulder and let fly, striking the nail with accuracy and power. Half a dozen such whacks were enough to drive the nail. It took me forty or so, like Lizzie Borden.

Willie John and I disagreed over whether or not to replace some rails I felt were too rotten. "No, let's get the stuff that's worse," he said, and we certainly had enough of that.

The job took about seven hours, less about forty minutes we spent sitting in the shade of the maroon pickup, eating lunch and listening to Paul Harvey, which was a pleasant break. I let Spook,

Willie John's portly cowdog, talk me out of a little salami. I tried not to let Paul lull me to sleep with his mellifluous tones and cornball timing, especially after I watched Jerry and Willie both doze off; I had an image of Bill flying over and seeing the three of us sprawled out motionless. And I looked out into the vastness of the Big Field and hoped I'd be ahorseback when we gathered it. I spent the last hour of the job stacking all the broken panels, posts, rails and wire into a big trash heap ready for burning.

A week later we branded at Rock Springs. The trap worked fine, but the day had other surprises in store. I helped Keith get the branding truck loaded up, then rode out with him. When I got there, Bill, already aboard Dusty, told me, "Grab a four-wheeler and go help on the drag."

I did that. It was a beautiful morning to be following a bunch of cows. Donnie Pettit, working near the lead of the herd, pulled up next to me as I headed to the drag and said, "Having fun?"

"Yep, you?" His smile and nod told me he was feeling pretty good about things today, too. "Don't push them too hard, or they'll bunch up and even try to go back," Donnie told me. "They know what they're doing. Let them go at their own pace."

Keith came roaring up on a four-wheeler a few minutes later and said, "You've got to push these cows along or we'll never get there." I knew that Donnie and Keith were among the best hands anybody could have on a ranch, but that didn't mean they always agreed about how to do things.

We got the pairs from the Big Field into the trap, and sorted off the mothers, who waited impatiently and noisily outside the fence while we worked their offspring. The Zehntners were there, and Willard and June, and Ronnie and Kurt Burns. Nancy Walter, a pretty, dark-haired woman from a pioneer Meagher County family who went off to work in San Francisco as a bank examiner, then missed Montana too much and came home, roped some calves as well.

After I'd wrestled for a while, Bill came up and said, "Go with Pat. Grab a four-wheeler and head down through the Dry Gulch and bring up anything you see so we can gather them quick after lunch."

Bill had just bought three brand-new Honda four-wheelers. They were considerably bigger than the old Kawasakis, or "cause-me-to-walks," as some hands called them, and therefore less maneuverable, but sturdier. One of the new machines had been given to Bergan at Lingshire, and Keith rode one, and the other Bill usually rode. I figured Bill was busy branding, so I might as well take his.

Off we went, Pat Bergan and I, zipping through the Dry Gulch side by side in our brand-new hot-rod Hondas like Hopper and Fonda, warm wind spilling the smell of sage and budding grass all around us. We rode without speaking for probably a quarter of an hour, dodging rocks and badger holes, over hills and through draws and coulees, the west fenceline always in sight.

Finally Pat slowed and put his machine into a gentle turn, ending up facing to the east. I pulled up next to him. "We should split up. I'll take high and you can take the low ground. I'm going to go up through those hills and see what I can collect. You could drive along the creek, through the bottom, and scare 'em out of there. I'll meet you down by the crossing."

"Okay, Pat," I said, and he nodded his white Stetson companionably at me and zoomed off.

The bottomland around Thompson Creek was muddy and rough, with tall grass disguising big frost heaves—known unabashedly around the ranch as "niggerheads"—that could high-center a four-wheeler in an instant. I skirted the very bottom, trying wherever I could to stay on firm ground, and made a lot of noise: "Hi! Hi! Get out of there! Come on, get up, get up, let's go!"

I felt a little silly, yelling where I could see nothing but green grass, brush and sparkling, fast-moving water—until half a dozen cows and their calves bolted out of the brush not twenty yards in front of me.

I tried to force them up the canyon wall. They turned up, then back down into the trees near the creek. I got off the four-wheeler and busted through the brush on foot, getting soaked, but it worked. I got the pairs up and out of there. They went up the left side of the canyon, and I got back on the four-wheeler and followed them onto a little sidehill trail. It wasn't much, just wide enough for a couple of cows, and that meant just barely wide enough for my machine. I kept close after them, tryng to keep them from running back down into the creek. The little trail gained some elevation as it curled around the sides of several hills by the creek—and then, almost without warning, it pinched down to nothing, leaving no way to get through the canyon and out to the other side, where the creek ran through much more level ground. I jammed on the brakes, worried that I was going to flip over sidehill; the pitch was very steep and there was just no place to go. I got stopped, turned the vehicle off and dismounted. If I'd been going any faster I would have been down the ravine and into the creek. The cows had no such problem. They went past the little spot on into the flat ground, thankfully, where I was trying to take them.

I was in a real mess. The four-wheeler was listing sharply downhill; it looked as though another couple of degrees of tilt would send it rolling out of sight. The rear wheel on the downhill side had nothing underneath it. Going forward was out of the question; backing it up didn't seem likely to work, either. If I could physically pick it up and drag it back a few feet, I could probably reverse out of trouble. I could have done it with one of the Kawasakis, but not with this new Honda. The boss's new Honda, at that.

I looked at it unhappily. I managed to shift the back end about six inches closer to solid ground, but pulling it uphill was pretty difficult. I would need a hand to get it out of there safely, and even then it was going to be tough.

I had no radio with me. I resigned myself to walking the couple of miles back to the branding trap and taking the grief I'd get when I arrived. But about ten minutes after I started, I saw Pat high on a hillside above me. I yelled, and he picked his way gingerly down the draw.

When I explained what had happened, he just nodded and said "Hop on." He drove most of the way back, then parked, and we walked up the hill to the spot. I was glad to see the four-wheeler hadn't tumbled into the creek.

Pat looked at it and shook his head. "That's what I'd call a precarious situation," he said. "Let's wrestle it uphill a little bit first. These damned new buggies are heavy."

We both lifted and pulled, and gained about a foot on the back end. "Now lift the front end while I try to back out of there," he said, and after a few false starts he got the buggy jockeyed back about ten feet. "You can drive it out now, right up through that cut," he said.

"Thanks, Pat."

"Yup," he said. "Be careful." And with that he walked back to his buggy and was gone.

I spent the next hour rounding up strays. Carefully. Then Keith hailed me from another four-wheeler and told me to follow him. He led me back to the trap, where Julia had lunch ready.

I had no idea if Pat had told Bill about my mishap, so I was prepared for the royal ass-chewing I deserved. But all Bill said was "You have a lot of nerve, taking off on my four-wheeler." I decided there was not much to say to that. I stole a glance at Pat, who had the faintest ghost of a smile on his lips. I should have known better; tale-bearing would be beneath Pat Bergan.

I wrestled a few calves in the afternoon, and then Bill put me on a vaccine gun, and I gave the rest of the calves pasturella injections. On the way back, we stopped to help Pat fix some fence, but he had just finished. "That ought to stay pretty sanitary for a while," he said with a smile. "I'll see you guys later."

When we got back, I helped Keith unload the branding stuff again. "Hey, Keith?"

"Yep?"

"How much are those new four-wheelers worth?"

"About five grand a pop."

"Jesus."

"Don't be driving Bill's anymore," he told me sharply, as though he was reading my mind. Still abjectly grateful for my good fortune and Pat Bergan's assistance and discretion, I headed out to change my water.

A couple of weeks later we were back in the Dry Gulch, Bill, Donnie, Keith, Willie, Tyson, Jerry and me, gathering pairs to move them up to the Dry Range, where they would spend the summer. As the four-wheeler rookie, I rated only the battered blue Kawasaki. Jerry was on the red one; Tyson on a motorcycle, Willie John and Keith on the Hondas, and Donnie on his ancient Polaris.

The morning was cool and misty. I humped and bumped along Thompson Creek and flushed out two dozen pairs, then drove them uphill and out onto the flats. I pushed them along until I got to a larger group being marshaled by Willie John and Tyson. Tyson headed back down toward the brush on his motorcycle, and Willie and I kept with the main bunch, picking off a few stragglers on the nearby hillsides.

Now Keith came zooming up. "Come on, both of you don't need to be pushing these. Willie, go check those coulees over there." He pointed to the east. "David, keep these moving. Where are Jerry and Tyson?"

"Tyson went back down toward the creek. I haven't seen Jerry in a while," Willie said. Keith shook his head and pulled his radio out. "Jerry, can you hear me?" There was no response.

It wasn't long until we found out why. I had maybe a hundred head now, and before long I met up with Tyson, who was pushing another two dozen. In a few minutes, Jerry came walking up to us. He was limping a little, and he had a deep scratch on his arm and another high on his cheek.

"What happened?" Tyson said.

"I was chasing these cows along this trail on the side of a hill, and the trail just disappeared and I rolled the four-wheeler down the hill," Jerry said, and I thought, it couldn't be.

Tyson said, "Down by the creek?"

"Yep," Jerry said. "The four-wheeler's in the creek now. I don't think it's too bad, but I couldn't get it out of there."

"I saw a flash of something red over that way," Tyson said, shaking his head. "At first it looked like a four-wheeler rolling down a hill, but then I didn't see or hear anything else and I thought, no, I must have been imagining things."

"Was it right before the creek comes out of the canyon, where it levels out?" I asked.

"Yeah, exactly," Jerry said. "The trail just disappeared and there was nothing I could do."

"Were you going pretty fast?"

Jerry nodded. "I was trying to get around some cows. They were headed back down into the brush."

I thought about how close I had come to doing the same thing. It really was a nasty spot. If I'd been going any faster, I would have been toast. I figured Jerry had been going a lot faster.

When we met Keith, pushing a group almost as large as the one I had, he glared at Jerry. "Get on that motorcycle," he said. "Tyson, take David's four-wheeler. We've got to get these up the hill and Bill's not happy at how long it's taking."

So I was afoot the rest of the way, running to cut off stragglers, cows looking for their calves, and calves who wanted to go back to where they were last with mom. We funneled them through a brushy draw, around a corner and through a gate, tough country for four-wheelers because it was so narrow. Being afoot enabled me to move through the tight spots easily. By the time we'd gotten up to a gate at the top of the pasture where Bill was waiting, I was leg-weary but very happy not to be Jerry, who was explaining to Bill what had happened.

Bill, considering he wasn't happy with the way we'd moved the cattle and was in a less than expansive mood anyway, was pretty restrained in his response. But that was before several calves got loose as we were pushing the cows through the gate.

His face darkened. "Goddammit, what's the matter with you guys? Go get those calves."

Before he'd even said that, Donnie, Tyson and I had jumped on four-wheelers. The calves were all but impossible for one person to herd in; they cut too quickly. Donnie roped one and put him on the back of my four-wheeler, and told Jerry, "Sit on him." But the calf got his legs loose, kicked me square in the back, and fell off the four-wheeler. I jumped off and landed on top of him—and Tyson, who had done the same thing. Finally, driving with the calf on my lap, Jerry holding on to his legs behind me, we got him through the gate. Bill was shaking his head in disgust, and, I suspected, trying not to laugh.

We got the other two a little more easily. When we were done, Bill said quietly, "That was a sorry excuse for moving cattle. You guys need to do better than that, and quit tearing up four-wheelers." He turned and walked to his pickup, and I knew he was right. Somehow it felt worse that he hadn't absolutely lost it with us, because I knew he would have been well justified.

That afternoon Tyson, Willie John, Jerry and I took on the ugliest fencing job yet—repairing a fence across Little Birch Creek, which was roaring with snowmelt. The line we had to fence ran down a steep canyon to the creek, then back up the other side. The canyon wall was covered with loose shale, and so steep that you could barely keep from sliding down to the water without hanging on to a fence post.

Once we finished the slope, the fun started. The remnants of the old fence had fallen into the creek, and debris had caught in it. It all needed to be cleared. The current was very fast, and the creek was much deeper than it looked. I waded in to grab a couple of large logs that were jammed into the thatch of smaller sticks and immediately went in over the tops of my hip boots, which made it a lot harder to maneuver, not to mention cold as hell. Tyson, Jerry and I got the large obstructions cleared and then poked and whacked at the little stuff while trying not to tip over. I got out, emptied my boots and went up to the truck for more supplies. Willie was starting on the creek crossing and needed the large spool of wire, a chain saw and two large eight-foot rails. Carrying it all across the creek was challenging in itself.

A large cottonwood at the edge of the creek served as a corner post. I took two wraps of wire around the tree, tied off and handed the wire out to Willie John, who strung it across to the fence post on the other side of the creek, stretched it and tied it off. I hammered staples into the tree to keep the wire from backing loose. We repeated the process twice, so we had three taut strands of wire across the creek. Then Willie cut the rails down a little with the chain saw, cut little notches around the rails a few inches from each end, then wrapped wire around them. We strung them on the fence close to the waterline, tying the rail wires to the strands of fence. Rails tied in a fence across a creek like this are called dead men and are effective in keeping cattle from crossing a fenceline in the water.

In order to finish the rest of the fenceline on the other side, I had to pack steel posts and a pounder across the creek. By the time we finished and packed everything up the canyon slope back to the truck, it was after seven. "Meet me back at the shop after you change your water, and hurry," Willie John told the rest of us. "We're going to go get that four-wheeler."

I was worried about getting my ten dams finished in time, but everybody had a lot to do, and I didn't hold up the party. We all got back at about the same time, nearly nine. We took the old maroon pickup—what else?—on the rescue expedition into the Dry Gulch. Jerry had no trouble directing us to the spot. It was indeed the precise place where I'd had trouble. We parked on the top of the hill above and went down to inspect what was left of the red Kawasaki. It wasn't a pretty sight. It had rolled a good hundred feet down the hillside. The handlebars and the lights were bent, but the front didn't look nearly as bad as the back. Both rear tires were flat; the rims were bent and the back axle was snapped.

"Jesus Christ," Willie said. "How long were you on it?"

"I got thrown off on the first roll," Jerry said. "I was lucky to get clear of it."

"I guess you were," Tyson said. "This thing is totaled."

"It still runs, though," Jerry said defensively. "See?" He switched on the key and punched the starter, and sure enough, it started up.

"Big deal," Willie snorted. "It sure isn't going anywhere on its own."

"Yeah, I know."

We strung together four catch ropes, and even then it was just enough to stretch from the front axle of the four-wheeler to the hitch on the Ford. The rope stretched tight, and the pickup's rear tires began to spin. "We're going to break it!" Tyson yelled. "Wait."

Tyson, Jerry and I lifted up on the four-wheeler as Willie pulled, and grudgingly it began to ascend. We had to push it all the way up while Willie pulled, but we got it up the hill and into the back of the truck.

"Thank God it isn't one of the Hondas," I said.

"We couldn't have done it," Willie said. "We'd have had to get the loader."

"Besides," Tyson said, "if it had been one of the new Hondas we'd all be dead anyway."

I rode in the back with the dead four-wheeler on the way back out. The sunset gave everything a pink-orange glow—hills, antelopes, gophers and sage grouse. If I hadn't been so exhausted, I probably would have appreciated it more.

A few days later, Bill came striding into the shop, quite pleasantly turned to Willie John and said, "Goddammit, Mr. Bernhardt, you need to pay more attention to your cowboying. I just flew Section 30 and you left a cow in there."

"Okay, Bill."

"Get her out of there. This afternoon, take one of these guys and go get those crippled pairs we left at Rock Springs and take them to Little Birch."

It was standard procedure, moving cattle, to leave sore-footed or injured animals behind so they wouldn't hold up the trailing. "One of these guys" turned out to be me. Willie and I hitched the white horse trailer to the crew cab, tossed in some fencing gear and headed for the Rock Springs branding trap. It was a blistering hot day, cloudless, and I noticed that the Big Field was a good deal drier than it had been at branding.

"You patch up the corral here and I'll go get 'em," Willie said, and so I did. The trap had taken a few hits during branding, and I repaired a couple of panels and moved one over to hike up a low spot. I had just finished when Willie John came back on the four-wheeler.

"What's the matter?" I asked.

"Nothing, but I've got a project for you," he said. "Hop on."

He took me maybe three miles, roughly northwest, until we came to three pairs. "I've got to go get the others," he said. "Walk these on in, would you? It'll save me a lot of time. Watch out for her." He pointed to a black-and-white cow. "She's nuts. I don't know what's the matter with her, but she keeps wandering off from the others. Her calf won't even go with her, he goes with the rest of them, but that red cow is the leader and she keeps wanting to veer north. If you'll get them as close as you can, I'll come back when I've got the others in the trap."

"Okay, sure," I said, happy to be able to play a role, even if it was a pedestrian one.

I was a little less happy in about twenty minutes. Keeping this wacky bunch headed in more or less the right direction was like trying to pour beer back into the bottle. I hiked them out a little south of where I estimated the straight line to the trap was, hoping to compensate for the north-pointing gyroscope in the red cow. But the black-and-white cow kept veering off at right angles. I'd cut over and get around her, but by the time I got her going okay the red cow and the rest of the group would be veering too far in the other direction. "Come on, girls, it's too hot for this," I pleaded, running to my left to keep in contact with the black-and-white cow. I could tell they were suffering from the heat, too. Their tongues were hanging out and their mouths were flecked with foam.

Finally I figured that I needed to get them to the fenceline as soon as I could, to help control them. That entailed going farther north than otherwise necessary, but that was all right with Red. It took almost an hour to get to the fence. These cows were in poor shape, and impaired to boot; I couldn't push them too hard.

The cows didn't like the fact that now they had fewer options. The black-and-white cow was particularly grumpy. But I had them within sight of the trap when Willie came to help.

"Good job," he said. "Thanks. It would have taken me a hell of a long time to get them all in together."

We got these six and the four Willie brought in into the trap, then into the horse trailer. We let them out in Little Birch and then it was time for more walking, this time in irrigation boots. Tyson and Wylie, who will run the swathers throughout haying season, were already cutting, so after I set my ten dams I helped Willie change Tyson's water on the Gurwell. As usual, the Gurwell was a pleasure cruise for me, with its rolling hayfields and forever views, a thunderstorm on the horizon and a pair of bald eagles cruising for dinner overhead. As I headed back to the truck after the last set, I figured I'd walked about twelve miles that day.

JULY 4

Back from irrigating, I asked Keith what was planned for the day. "I'll call Bill," he said. Unspoken was the question of whether or not we'd get any time off for the holiday. My guess was not; Bill was considering having Wylie and Tyson swath again today, and the basic rule was, if one member of the crew worked, we all worked. Willie came in and Keith said he needed to go do Tyson's water on the Gurwell again. Still in my boots, I went to help him.

"Shit!" Willie John spat disgust and Copenhagen out the window of the truck on our way back. "He couldn't give us a goddamned day off on the holiday."

The crew was hot, tired and haggard, and everybody got a little more grumpy when Bill called Keith to say that he wanted the swathers moved to the next hayfield today, even though it's illegal to haul a load on the highway wider than nine feet on a holiday. Bill has a stubborn streak; I knew that he wasn't about to be dictated to, either by state law or the crew's expectations, of which I was sure he was aware.

We moved the swathers, without blocking any holiday traffic, and at noon, Bill gave us the rest of the day off. That brightened the crew's mood considerably. I drove to Livingston for a Fourth of July barbecue. I tried to be sociable, festive even, but I kept falling asleep in the middle of conversations. People who don't work on a ranch have a hard time understanding this.

chapter 17 • pestilence and fire

AS HARD ON THE RANCH AS WINTER'S STORMS HAD BEEN, THE summer presented challenges almost biblical in their immensity. Plagues of grasshoppers and locusts were not unknown in Montana; in the twenties in eastern Montana, homesteaders reported grasshoppers so thick they ate the wooden handles off the tools. We had other problems at Birch Creek—pests both animal and vegetal, and the terrible specter of range fire.

One day in early July we pulled old maroon into the shop and Keith helped me load the 200-gallon spray tank with a devil's brew of water and two kinds of herbicide, 2,4D and Tordon. It was weed season.

In the last two decades non-native noxious weeds have invaded Montana with a vengeance. The pests are plentiful: Burdock. Canadian thistle. Dalmatian toadflax. Henbane. Ragwort. Sulphur cinquefoil. But by far the worst are knapweed and leafy spurge. State experts estimate that four and a half to five *million* acres of Montana are infested with spotted knapweed and half a million to eight hundred thousand acres are plagued with spurge. "Knapweed's a road and railway plant and spurge is a waterway plant," Park County weed-control expert Clyde Williams told me. "They're both incredibly hard to stop, and they choke out anything else growing."

Seedpods are the villain with knapweed: They are nearly indestructible, and they're easily carried—by cars, trains, people and animals. Spurge is even more versatile. It can spread from seeds, but also with rhizomes, or underground shoots, that can creep for several yards before reemerging.

Weed fighters are trying an assortment of biological weapons. Introducing flea beetles, which attack root systems, can help with spurge, and gallflies and root borers are employed against knapweed. Everyone, except perhaps the chemical companies, would love it if such natural weapons could solve the weed problem. "The insects are not a silver bullet," sighs Gallatin County weed control district director Dennis Hengel. "They're an added stress agent for the plants. And in concert with an aggressive spraying program, they can help. But the biological factors alone won't do it. With them you measure progress in feet, not acres."

Bill Galt had used the insect weapons with only limited success, so for the next few weeks Jerry and I would be gunning for knapweed and spurge with chemicals. As we filled the big tank, Keith told us, "Be careful how you mix this stuff. Always do it here in the shop where we can control it. Wash the end of that hose off after it's been in the tank. If the wind kicks up, don't spray. Every time you spray, record where you were and how much you sprayed and what the mix was in this notebook. And wear this protective gear." He handed us paper respiratory masks and disposable white coveralls.

Then Keith gave us a good primer on recognizing spurge and knapweed. Unfortunately, we didn't have to go far to find examples. The two corner stackyards were our first targets. It's important to keep the weeds out of the stackyards, because they can spread from there so easily, both with the hay through cows' digestive systems and with the vehicles entering and leaving. When we pulled into the larger of the two stackyards, Keith said, "There's knapweed. See, that spiky-looking plant. That part's brown because we sprayed it last year, but it didn't die. See the green shoots about to flower? And there's a few more by the fence, probably seeded from that first plant." We soaked the plants with herbicide, and a toxic odor welled up around us.

"One of you can drive around the outside of the stackyards with the truck, and the other one take the hose and nozzle and spray whatever weeds you see," Keith said. "But first, come with me. I want to show you some spurge."

We drove down near Little Birch, and Keith found a little hillock covered with bright yellow flowers. "That's spurge," Keith said. "There's a couple of other plants that look similar. If you're not sure which you've got, cut off a stalk. See that milky stuff that comes out? That's spurge. The other plants don't have that.

"Keep an eye out for thistle, too. Don't spray anything if you don't know what it is."

We spent the morning and most of the afternoon spraying stackyards. All of the stackyards on the bar had some knapweed; a couple of them had quite a bit.

Later in the afternoon, we headed down to the spurge patch Keith showed us near Little Birch and looked for more. We found some, but as we looked I was startled to realize the entire hillside I was tramping over was covered with knapweed. So we did some carpet bombing, laying down about fifty gallons of spray. It was sobering to see how little other vegetation remained on the knapweed-covered hillside.

After that, Jerry and I would spray weeds almost every day. Often conditions were best right after irrigating in the morning and just before changing water in the evening; the wind tended to be lower then. One morning about seven A.M. Jerry, his brother Joel and I took the truck back down to Little Birch to knock down another patch of spurge we'd spotted. Joel was a beefier version of Jerry, wide and muscular, good-natured with a slightly goofy sense of humor. He, too, had done ranch work all his life, and liked to come out to Birch Creek on his days off from the stockyard in Billings, and help Jerry with whatever he was doing.

Some of the spurge was beyond reach of the big sprayer, so I loaded up a backpack and walked the creek bank. Joel manned the hose reel, feeding hose to the big sprayer for Jerry.

I came back to get another load and froze. Joel was standing in the bed of the pickup, looking ahead to where Jerry was spraying spurge. "Hey, Joel," I said. "Turn around real slow."

He did so and almost jumped out of the bed. Four moose, including one very large bull and one only slightly smaller, were gathered around the back of the pickup, the nearest perhaps twenty feet away. Unlike the cow I'd surprised, they didn't seem angry, just curious about this unusual early-morning invasion of their territory.

They stayed for another few minutes. I was fascinated to see them this close—huge, ungainly but handsome, with a presence unlike any other animal in my experience. These creatures conveyed a power and an indescribable wildness. Of course I knew firsthand they could be dangerous, particularly protecting calves, but not as dangerous as humans have been to them, shrinking their habitat and hunting them by means legal and otherwise. This encounter engendered not fear but wonder. Another Birch Creek blessing to count.

We'd just finished, about five o'clock in the afternoon, when Bill called. "We've got another fire in the Dry Range. It's right next to the old fire—might be a rekindle, they're not sure. Change your water quick. Then Willie, take Jerry and David up there and help. Pat and Donnie will meet you there, and the Forest Service is sending a pumper."

There had been a fire on the Dry Range a week ago. I hadn't been involved in fighting that one, but every time I missed something, fate had a way of sending it around again for me.

When we got to Den Gulch, the smoke was heavy in the trees, and a few minutes later Willie brought the truck to a stop thirty feet upwind of a hillside crackling with flame. "We've got to get a line around that," Willie said sharply as we rolled to a stop. I grabbed a Pulaski, a firefighting tool with an axe blade on one side of the head and a flat, hoe-like implement on the other, headed for the downwind side and started hacking. The smoke was choking. I had wetted down my neckerchief with a canteen in the truck, and I raised it now over my mouth and nose.

Digging fire line meant furrowing the earth to remove anything that would burn—pine needles, downed timber, all vegetation—chopping low branches too so they couldn't lift the fire over the line. Creeping juniper and scrub pine made it tough going on the rocky hill. Trying to chop away a creeping juniper plant is like trying to cut through a steel-belted radial. I was hacking as fast as I could and making less progress than I wanted. Willie John, Jerry, Joel, Donnie and Pat were all spread out around the blaze, which was roughly a couple of a hundred yards long and half that wide. And getting wider. The wind was kicking up, and I was forced to retreat twice to get around new burns.

The next two hours were nothing but grunting, digging and swearing, but we gained on it. Donnie, Willie John and I got around the toughest part of it just as the Forest Service pumper showed up. They ran lines quickly and got the leading edge knocked down considerably, but they ran out of water in about fifteen minutes. Willie went with them to help them fill the pumper at the Den Gulch tank, and the rest of us kept slogging away. By the time the pumper returned, we had a solid line around it. Now it was a matter of cleaning up hot spots. We sprayed, shoveled, resprayed and reshoveled, cruising all over the fire zone. The wind had moderated quite a bit, it looked as though it might rain, and I leaned on my shovel for a few minutes, drank a cup of Pat Bergan's coffee and talked quietly with him, Willie and Donnie.

"Good we got around this pretty quick," Pat said. "If it had gotten down into that canyon it would have been a son of a bitch."

"Do you think this flared up from a hot spot?" I asked.

Donnie shook his head. "Shit, it was absolutely out when I was up here yesterday. I think we got another lightning strike this afternoon, in almost the same spot."

That made sense to me. The fire seemed to have started a little way away from the old burn. In fact, I thought we were fortunate the other burn was there, to help contain this one.

We spent another hour or so digging hot spots. Glowing coals were easy to see as dusk faded into night. The Forest Service fire boss said finally, "I think we're in good shape. Are any of you going to be up here tomorrow?"

"I'll check it in the morning and again in the afternoon," Donnie said. We rolled hose, slapped out a few more spots, gladly accepted box dinners from the Forest Service, and headed out of the forest to Birch Creek.

Back at the shop, Willie and I unrolled the hoses we'd used so they would dry out, and I helped him fix a flat on his pickup, then went into the bunkhouse and fell asleep before I could even stagger to the shower.

I hated spraying weeds, but I understood the necessity of it. But my next pest-control assignment was much worse—equally toxic and much harder to execute.

Bill brought five ten-pound bags of strychnine-laced oats into the shop one summer morning. "Keith, show David and Jerry how to put this stuff out on the wheel line at the Stevens. Make sure they're very careful with it. It'll kill anything that eats it."

So Keith showed us the logistics of sowing death around the huge field. You opened the ten-pound feed sack just on one corner, to make a small pour spout. You found a gopher hole—not hard, the field was covered with them—and you poured a few grains of the red-dyed poisoned oats into the edge of the hole—the gopher foyer, as it were—and went to the next. "That way you don't have to touch the stuff," Keith said. "Don't waste it; we don't want any lying around. Just a tiny bit, a teaspoon in each hole. That's it. Keep going until you've gotten all the gopher holes, or you run out of oats."

When gophers show up in profusion as the snow melts in the spring, the biggest shooting season in the state, this one with no regulations or limits, gets under way. Gopher hunting is probably the most prevalent recreational activity in the mountain West in the spring—one I've never particularly cared for. I've seen people use everything from a .22 to a shotgun to a one-ton truck to kill a gopher—it's not unusual around here to see someone swerve across two lanes of traffic to *hit* a gopher—but I'd never seen them poisoned, much less done it.

A wildlife biologist would tell you we have no gophers in Montana. They are actually Richardson's ground squirrels, and while they were not placed on this earth to provide a moving target for men who like to shoot things, it often seems that way in Montana. They are cute little animals, as well as a significant headache for farmers and ranchers. Their holes, like those of badgers, can break a cow's or a horse's leg. And of course where a gopher digs, grass isn't growing. And they are prolific breeders. Most of the time, on rangeland, a rancher grits his teeth and puts up with them. But on a cultivated field, they can make a huge mess, and they had done so on the wheel line.

Gopher Day was certainly one of the hottest days of the year, and there was not a cloud in the sky or a speck of shade on the ground. I tried to be methodical, walking off squares of territory, then covering everything in the square before moving on, but even with the attempt to impose order, the whole thing felt surreal: wandering sun-blind uphill and down, carrying a feed sack, bending every few feet, lightening the load one teaspoon of death at a time. Unless you counted mosquitoes or cockroaches, I had never deliberately killed hundreds of sentient beings before, and by two in the afternoon I was sunburned, sore and disgusted with the whole operation. We still had two bags of oats to spread and almost half the field to cover. The job took eight hours.

Bill Galt has a finely developed feel for when his hands have reached the snapping point. So in mid-August, he told Keith to take off for a week.

Willie John was gone for a while for a different reason: As we were working at the Stevens the other day, Julia called on the radio to say that Erin had gone into labor, and Bill told Willie, "Get out of here and take as long as you need. Let me know what's going on." Nobody had heard from Willie since, but we did know that Erin had had her baby, a nine-pound girl named Marisa, and that everybody was fine. Jerry, meanwhile, was off getting married. So it was Bill and me and Donnie on the Stevens and the Dry Range, and Pat Bergan at Lingshire, and that was it.

A few minutes after one that first afternoon, Dan Hurwitz called on the radio. "Bill, the Forest Service wants a Cat up on a new fire somewhere near Kings Hill. Are you willing to part with one?"

"Well, I'd rather leave them close to home in this weather," Bill replied. "Let me call the Forest Service."

A few minutes later I heard him call back. "Well, they seem to want it real bad. Actually, they want two. If I haul one of mine up there, will you take the other one? Keith's off and I don't have another driver."

"Sure," Dan Hurwitz radioed back.

The D-8 was at the Stevens, and Dan made plans with Donnie Pettit to go load it. Meanwhile I helped Bill hook the machinery trailer to the yellow Kenworth and load the D-6 on it. He left for the Forest Service office in White Sulphur, and I found a set of lights in the shop to hook up on one of the Cats, filled a two-gallon jug with coolant, and took Bill's truck in to meet him there.

And there we stayed, for a hell of a long time considering there was a fire tearing up a mountain.

Fire has always been the most dreaded of all disasters in the northern Rockies. Joseph Kinsey Howard wrote of one mammoth grass fire in the 1890s that swept down out of southern Saskatchewan, leaped the mighty Milk River, already forty miles into Montana, and ran southward another thirty miles before burning itself out in the Little Rockies. Norman Maclean's "Young Men and Fire" gave me a perspective on how puny my firefighting efforts had been thus far. I had seen the aftermath of the great Yellowstone Park fires of 1988 and had walked through a desolate fire-ravaged area in the Jemez Mountains of New Mexico. Neither are sights I will ever forget. In the 1990s West, though, firefighting was big business, and that meant big bureaucracy.

The little White Sulphur Springs Forest Service office was frazzled but efficient. There were a zillion logistical details to tend to: smoke jumpers to meet at the airport, campers who couldn't get back to their campsites because of roadblocks, more supplies and firefighters to summon from all directions. And a couple of Cats to get to the fire.

The problem lay farther up the line. Before the fire could be fought, the paperwork machine had to be fed. First the Cats and the trucks had to be inspected as to condition, then all the specifications of each machine had to be noted, in addition to the ownership and registration of the trucks and trailers.

As I waited in Donnie's wife Rose Pettit's office for the Great Falls Forest Service office to fax hauling and equipment usage contracts, I noted the afternoon fire weather forecast from the National Weather Service office in Billings, posted on a bulletin board. "Red Flag Warning for All Zones," the report read. "The upper level ridge continues to dominate the Northern Rockies. Temps across the district this afternoon are 95–100 degrees and many locations have humidities less than 10 percent. A weak Pacific cool front will enter Western zones later tonight, with increased winds along the front range the primary red-flag mechanism. Any thunderstorm activity will be high-based and dry. The front will move into the western Dakotas by Monday afternoon, with gusty winds expected to blow across the entire district all day. The winds will diminish Monday evening, and then it's back to waiting for the next trigger to set off another red-flag situation. At this time it looks like the hot dry weather will be here through this upcoming week, with temps in the 90s and afternoon humidities flirting with the 10 percent plateau. Summer in the Northern Rockies . . . what a great place to live!" Well, actually, yes. But not without periodic trials, by ice, by fire, by bureaucracy.

When the contracts finally arrived by fax at five-thirty, another problem presented itself: The contract for the use of the Cats was satisfactory, but instead of the customary rate of about $80 an hour for hauling the Cats up to the fire, the Forest Service was proposing to pay Bill and Dan a rather niggardly $3.80 a mile, far less than they would have taken for an over-the-road hauling job, much less this sort of a situation, involving driving extremely heavy loads over tortuous dirt roads, into a situation uncertain at best.

Bill politely protested and was eventually referred to a new supervisor in the Billings Forest Service office, who told him the contracts conformed to the latest Forest Service regulations. Again politely, in a conversational tone, he replied, "I don't care. Nobody will haul for that rate, and I sure won't. As a matter of fact, without getting anybody's hair up, I think I'll just take these machines and go home."

I heard a squawk on the other end, which he cut off. "The last fire I sent equipment to, I billed you twenty thousand dollars and got paid eight, which didn't even cover what I'd had to pay my operators. I swore I was done with you then, but here we go again." Finally they came to a tentative agreement, based on the rather unrealistic premise that Bill and Dan would simply unload the Cats where the Forest Service wanted them and leave immediately. "I've never taken equipment to a fire where the supervisor released the trucks right away," Bill told her. "They always want the trucks and drivers to stand by so they can get the Cats out quickly if they need to." If that happened, she said, the supervisor at the scene would have to call her for permission, and at that time further terms would be arranged.

The local firefighters were apologetic and more than a little worried. Reports from the fire indicated it would be a long fight. Apparently started accidentally by a logging crew, the blaze had already spread over some 700 acres of dense pine forest, helped along by extremely hot, dry air and a wind of ten to fifteen miles an hour.

When we finally got rolling, I led the way in Bill's pickup, flagging for the two eighteen-wheelers. Donnie had met us in town and had given me a flashing light that plugged into the cigarette lighter and supposedly stayed on the roof of the truck with a suction pad. It didn't and I had to drive with one hand and hold the light on the roof with the other. We hadn't gone far before Dan Hurwitz radioed: "I've got to pull over; I'm losing power and my truck's making a heck of a racket." Apparently, his turbo unit was failing. Bill went ahead, with Donnie flagging for him, and I stayed to help Dan and backup trucker Dave Collins with the ticklish job of changing out trucks. The trailer carrying the D-8, all 80,000 pounds of it, was sinking dangerously into the asphalt of the shoulder of the road. But we managed to get the switch made and get back on the road in about half an hour.

The instructions I got were to turn off the highway to Great Falls just before Kings Hill and follow the road to the Sheep Creek campground. It took us a little over an hour to reach the Forest Service roadblock, and another fifteen minutes to the staging area where they wanted the Cats unloaded.

It was sunset, and the pall of smoke had turned the sky from orange to a dull, unearthly yellow. The mountain, though, was orange—a huge bloom of flame visible dead ahead, growing visibly. Smoke poured in enormous clouds from the fire we could see, which was only one edge of the blaze. I had never seen anything remotely like this. It made the Ellis Canyon blaze I'd helped to fight look like a campfire.

I helped Bill and Dave loosen the chains and get the huge Cat unloaded. Chuck Swanson was there to run the D-8; another expert White Sulphur Cat skinner, Mark Herzog, was already getting the D-6 ready to roll.

Bill Galt and Donnie Pettit and I walked in behind the Cats, seeing where they were headed. We went probably half a mile to a clearing on the edge of a bluff. The fire was just across the draw, an utterly amazing sight. We could see it eat trees, one after another. "Look at that son of a bitch," Bill said. "It's so hot it's making its own wind, feeding on itself." As he pointed, I saw the flames at the fire's leading edge churning in a whirling motion, almost like a tornado of fire, something beyond Dante, or the Book of Revelations. The heat was fearsome, even from here.

A firefighter came up to us and said, "Please wait here. There's a supervisor, John Metreone, coming up here and he wants to talk to you." In about five minutes, Metreone, his face streaked with soot and sweat, came up to us, took off his yellow helmet and sighed. "We're going to have to pull back," he said wearily. "We can't stop it here. It's going too damned fast."

Bill said, "You know, I'm under orders from the Forest Service in Billings to drop these Cats off and let you move them wherever you want them."

Metreone shook his head emphatically. "No, that won't work.

We're going to pull back clear to the highway. We'll need you to take them out on the trucks. I'm not sure where we're going back in, but it'll be a ways north of here."

"Well, you'd better get on the radio to Billings and get this lined out," Bill said. "I want the hourly rate I've gotten before, not the per-mile, starting at three o'clock this afternoon, and I'm not going to negotiate."

Once the Billings office understood the situation, it was an unconditional surrender. Bill shook his head and said to me, "All that wasted time. I'd damn well better get paid without any nonsense, or this will be the last time I send them any machinery."

So Mark and Swanny walked the Cats back out, and we followed, the smoke thick on us and the hot breath and flat thunderous roar of the beast at our backs. It was dark by the time we got the big iron back on the trailers and headed back to the highway. We were stopped at Kings Hill, where the Forest Service had set up a staging area, waiting for the strategizers to decide where to attack what was now known as the Coyote Creek Fire.

"We'll be up here until October on this one," one firefighter said to me as I waited with Bill's pickup for both of the big diesels to pull on up the grade. "We're looking at forty, maybe fifty thousand acres." I wondered, if he knew all that, what he was doing directing traffic, particularly since as far as I could tell, not so much as one Pulaski had been raised in anger against this fire since it was reported eight hours ago.

When the trucks arrived, we took the opportunity to rig the D-8 with the set of lights. That took five minutes. Then we waited another hour and a half before we got instructions to move out. Right in front of us was a school bus containing a firefighting crew from the Blackfeet reservation, and we followed them in.

The fire seemed unreal now, like a jagged piece of orange construction paper glued against the black. It was no longer possible to see what was at risk—the trees and animals, the cabins, even the nearby ski area. The forest was visible only where it was being consumed.

We stopped where they wanted us, and the Blackfeet filed out of the bus silently, wearing yellow hats with miner's lights, carrying axes, wearing bulky packs, some with oxygen tanks. I thought of that snarling orange monster I'd seen this afternoon, and I was ashamed, not so much at my puny efforts at the dump, at the CRP burn, at Ellis Canyon, but ashamed to be glad I was not in that quiet procession trudging into hell. I wished them a safe return, and Mark and Swanny, too, who had their own corner of the inferno to confront. We put those Cats down and got out of there then, Bill driving the Kenworth and Donnie riding with me, and the bunkhouse looked pretty damned good, at three A.M.

I was an irrigating robot three hours later. Thankfully, everything went well and I was back in the shop, risking an extra cup of coffee, by seven A.M. I barely got away with it, draining it as Bill roared up. "Give me a ride back to the house, would you, and then clean and service this truck?"

The service job on Bill's ranch pickup went pretty easily until it came time to clean the interior. Bill's truck was always jam-packed with all kinds of stuff. I decided I had to take everything out, clean and then reorganize. "Everything" turned out to include an astonishing fourteen coats; twelve caps; one straw Stetson; two pairs of coveralls; chaps; a bridle; two sorting sticks; two boxes of ammo; a box of fireworks (which truly puzzled me until I found out they were Wylie's); a box of groceries, which included a loaf of bread, half a package of frighteningly elderly salami, half a jar of similarly lethal mayonnaise, and two donuts; three cameras, five boxes of unexposed film and two rolls already shot; an empty cooler; and thirty-seven warm cans of Pepsi.

I put most of it back, repossessing one of the coats, which was mine; putting a cold-pack in the cooler along with six of the Pepsis; and taking the liberty of disposing of the salami and mayo. Then I hooked up the power-washer, backed the truck out onto the driveway and did the outside.

I kept an ear cocked for the radio, seeing if I could hear any word of our Cat skinners, and around two, Donnie came on the radio, relaying a message for Bill: They wanted to stay on for another twelve-hour shift without being relieved.

"Well, that's okay," Bill replied. "But tell them to shut it down if they get too tired and we'll get somebody else up there."

I shook my head. I could understand them wanting to make the most of it while they were up there, but as tired as I was this morning, I couldn't imagine how they must feel, pulling levers all night in the heat trying to get a line around the worst of it, and then asking for more.

Chuck Swanson came back to the bunkhouse every other day for some rest between marathon shifts running the D-8 at the fire. Despite the dire predictions, after ten days the work was nearly done. With a little assist from the weather, the massive efforts of hundreds of firefighters had paid off. The fire had covered only about 3,500 acres—considerable, surely, but far from what many were predicting in the early hours of the blaze.

One evening Swanny ate dinner with me, and regaled me with cowboying stories. He grew up poor and quickly in the Pryor Mountains of eastern Montana, and took his first riding job at age fourteen, for the wage of two dollars a day. When payday came, the rancher took him into town and bought him a pair of boots instead of paying him.

"He thought I didn't know how much them boots cost," he said. "Shit, I knew them goddamn things cost twenty-two fifty and he owed me sixty. But I needed the job and I needed the boots."

Along the way he became one of the best bronc riders of his time, appearing in the National Finals Rodeo several times. Now he's fifty-five and looks forty. He's done well buying property and breeding racehorses. His daughter, Molly, is one of the world's leading professional barrel racers—fourth at the moment—and he still judges rodeos when he's not working around his place or away on jobs like this one. He first hired on at Birch Creek in 1971, for Bill Loney, and worked there several times over the years.

He is a completely self-made man, both a champion and a victim of his culture, deeply xenophobic, homophobic, everything phobic, suspicious of everything different from the Montana ranch life he loves, but one of the hardest and handiest workers ever to pull on boots. He enjoyed working for Bill, he said, but he missed the days when Birch Creek was pretty much a horse outfit.

"All them machines," he said with a wave of his arm toward the shop. "They're okay, I guess, if a man can afford them, like Billy. Hell, I can run 'em, but I liked it better when we done it all ahorseback." I knew that was a prevalent sentiment, and as much as I'd rather be riding than four-wheeling most of the time, I knew that Bill had his enormous investment in machinery for a reason. He'd been here when it was all done with horses, and the ranch was not as productive and took a lot more work than it does today.

Sometimes I felt closer to the other hands than I'd ever felt to coworkers. Hard physical labor, working together to get things accomplished, and protecting each other from harm has a tendency to do that. But other times, I felt more isolated than I had in my life. I was not like them, after all; I didn't grow up like Swanny or like Bill Galt, for that matter. And my work life had been as different as it could be.

Also, ranching is almost completely a monoculture, and the casual racism so frequent in conversation still jarred me. Indians were "blanket-asses," frost heaves "niggerheads." (Willie John Bernhardt was part Indian, and one day I asked him if it didn't bother him when somebody told a racist joke about Indians or even poked fun at him. "I've got more important things to be pissed off about," he said with a grim smile.)

It was perplexing to me. These men I worked with were the fairest, most honest I'd ever known. They'd judged me not on my age or my bookish, sedentary background but on my ability to do the job and the effort I put into it. What if I'd been black? I knew the answer: I might have been given a chance, but I wouldn't have been nearly as welcome. Sure, there are some black cowboys; always have been. But damned few, and few of those are in Montana. I wondered how it would have been in those days ahorseback that Swanny remembered, or back in the homestead times a generation or two before that. Of course, racism was then a nationwide fact of life, and while I was considering this cold-eyed, it didn't hurt to remember that quite a few early settlers in Montana were refugees from the Confederacy.

AUGUST 8

I sprayed weeds along Little Birch Creek with a backpack sprayer today, carrying five gallons of poison on my back. A gravity-feed nozzle system of dubious design delivered a dribble of chemical more or less where I pointed it. The spurge had spread along both banks of the creek like a bright yellow cancer, choking out grasses, lupine and iris still trying to bloom beneath it.

As I squirted Tordon on it sparingly, avoiding at least the surface of the creek, I noticed a rustling in the high brush to my right. I watched carefully as I walked, and pretty soon I saw what was causing the noise. It was Super Chicken, the half-tame pheasant, and he cocked a bright gold eye at me and let me walk up to him and touch his resplendent tail feathers before waddling a few paces to a safer distance.

I didn't want him ingesting any more of this herbicide than necessary, and I tried to shoo him away, but he followed me for a good two hours, occasionally pecking angrily at my ankles. Perhaps it was my own deep concern about the chemical, but I thought he was trying to get me to stop spraying. After all, he was a surviving non-native species himself. I will always be haunted by the fear that I harmed him on that beautiful and still summer afternoon.

chapter 18 • leaf-cutter ants

WE STARTED HAYING JUST AFTER THE FOURTH OF JULY, NOT on our own fields, which weren't quite ready, but on Phil Rostad's ranch out by the nine-mile Y, south of town. Bill had bought the hay "on the stump" for about twenty-five dollars a ton, half what it would be worth cut and baled.

Bill said, "Climb in the tractor with me and I'll show you how this new baler works." I found out what the mass of cables and wires running between the baler and the tractor were for. The onboard computer told Bill which way to steer to get maximum coverage of the windrow of hay left by the swathers. It told him various other information about how the baler was working—most important, at what pressure the bales were being formed. That in turn let Bill know if the hay was dry enough to bale or still too green. If it was baled green, the hay would mold, like a few of the raunchy bales I'd fed this winter.

Time to go to work. Bill stopped and let me out of the tractor, and as I hit the ground Keith said, "Come here, I want to show you something." He climbed into the Dresser loader. "Hop up here and hang on." There wasn't enough room for me in the tiny cab, so I climbed onto the step and held on to the grab rail outside the door. "Here's your transmission lever. You've got three forward speeds. It's geared super low. Don't over-rev it. This lever controls the pitch of the forks"—he pushed it forward, and the forks tipped down, and he pulled it toward him, and they angled upward—"and this one takes 'em up and down." He demonstrated.

"That's it. You put one bale on top of another, good and square, and then pick up the two and bring them to where we're stacking. Make sure the strings are on the sides, not on the top. Sometimes the baler spits 'em out and they roll on their sides. Don't go too fast with it. Be careful. Now go stack hay."

I'd admired Keith's technique at running this loader, and I'd figured that was one thing I'd never get to do. My first few attempts were ridiculous. I didn't get the forks low enough and ran them into the bale. Then I got them too low and dug up the field. I dropped three or four bales trying to stack them straight.

It was a lot harder work than I expected, intense mentally and physically too. The Dresser had basically no suspension, and the slightest bump was jarring, causing me to tense and brace constantly to avoid banging into one side of the cab or the other.

Within an hour I had a basic understanding of it, and I started to make progress. It was exhilarating. I was eight years old again, playing with toy trucks in the dirt. We worked until ten and got 216 bales out of the field and into the corner stack. We hauled them with the yellow Kenworth, loading them 54 at a time on the lowboy trailer.

As soon as we had a load ready, Keith would take the loader from me and stack the bales on the trailers. Then he gave me another job. Each row of bales was secured by a thick nylon web strap. The straps were rolled up, and I had to hang on to one end and throw the roll as hard as I could over the stack of bales so that it could be reached on the other side. Then I would thread it through a catch and ratchet it snug using the cheater bar that I knew was always stowed under the sleeper in the truck. It had a hook on one end to turn the ratchets. Keith or whoever else was catching the straps on the other side would do the same thing.

Bill was forever hammering on safety, and he took this occasion to hit a few licks. "Be damned careful around the trucks," he said. "Somebody gets killed every year by falling hay bales. When you're strapping, or when Keith's stacking, watch what's going on." The trick was to throw the strap so it would unfurl in the right place, and if the wind was blowing you had to allow for it. A miss meant you had to pull it back and re-roll it before you could try again, which was time-consuming, but after the first couple of tries, I didn't miss much.

The next day was a dawn-to-dark hay day. We were at Rostad's by seven A.M. "Gas that loader up and get to stacking," Keith told me. Jerry was operating the Chevy loader, and both Kenworths were running, Keith driving one and Dave Twitchell, a driver from White Sulphur, the other. Jerry and I stayed ahead of the trucks all day, and we got 500 bales trucked out. I was getting quite a bit handier on the Dresser. It's not good for driving long distances to get bales. It's too slow because of the way it's geared. But it's much more maneuverable and quicker in confined spaces. So I would get the bales closest to the place we set up to stack, and Jerry would range farther afield with the Chevy loader.

It was a great feeling, getting quite a bit accomplished, but the wind made it harder than yesterday. Hay flakes and dust were blowing all day, and everybody had red, swollen eyes.

"I hate haying season," Keith said. "The same thing, over and over and over." I could see what he meant, but it was still new to me, and I was unabashedly enjoying it. We were driving back to Rostad's in the blue Kenworth. We still had more than 200 bales to pick up there, and Keith said we had another little stacking job that afternoon. We whipped through the bales at Rostad's—having two trucks going really helped—and after lunch we moved the loaders over to a little hayfield right in White Sulphur Springs, near the golf course. Bill and Gary Welch had baled it yesterday, just over 200 bales. This field, formerly owned by the Doig family, which includes the illustrious novelist Ivan Doig, is now owned by a local doctor named Dan Gebhardt, who usually sells the hay to Bill. It felt odd to be haying right in town, with kids going by on bicycles and Sunday golfers out hacking around the course next door. It did clear up one thing for me, though: The Mystery of the Golfing Heifers. It was easy to see how errant shots got into the Gebhardt field and into the hay Bill baled and bought.

Fourteen-hour haying days were the norm, it turned out. Often I would not leave the retriever once I got in it until we were done for the day and it was time to gas up—partly because the batteries were bad and I was never sure it would fire again if I turned it off, but mostly because I was busy.

After Rostads we hayed the fields I had irrigated up on the bar—the strip above the shop, and the fields farther down on the north side of the road. The runs from field to stackyard were short and we were able to get quite a bit of hay put up quickly. The tricky part was stacking the hay squarely and neatly so it could be easily retrieved in the winter, leaving room to get between the rows, but not so much that space was wasted. Unfortunately Jerry tended to set his loads down cocked a tad to the left, while my tendency was to err toward the right. This led to an explosion when Keith inspected the first couple of rows we'd stacked.

The first I knew of it was the sort of third-person radio traffic I hated the most. "Birch Creek One," Keith called exasperatedly. "Bill, these guys have this hay stacked so cockeyed I'm going to have to bring the loader over here and restack it two bales at a time. They've got it angled so it's cutting right through the middle of the stackyard."

"They" listened with chagrin as Bill said, "Well, damn it, we can't spare the time to keep fixing mistakes. They have to do it right and we don't have time for them to be learning. You've got to show them, Keith."

So Keith called us directly: "You guys bring those loads back to the stackyard and wait for me." He was pleasant enough then, and his advice was helpful. He showed me how to swing up past the stack, square up and back straight up to the bales, then lift the bed partway, then back the rest of the way, set down the load and drive straight back out on the same track, so it would be there to guide me back on the next load. After that we had much less difficulty.

Keith also gave me another task that I relished: taking one load from each land and weighing it at the scale shed. He showed me how to calibrate the big scale, then how to get an empty weight with the truck. When I weighed the loads I'd take a little card printed up for the purpose, put it in the slot and stamp it with the gross weight. Then I'd subtract the empty weight, and divide the remainder by ten to get the average bale weight. Usually it was right around 800 pounds, more if the hay was really green, less if it was really dry.

I'd first met Bill Galt in the scale shed as he was getting an empty weight on one of the Kenworths after hauling some hay he'd sold to a neighbor. Back then, in February, the big drive-on scale had been part of an utterly foreign environment. Now it was another piece of equipment that made sense to me.

Many of these hay days, I would get up early, make lunch and fill a water jug, feed the hounds and get running before six. If there was still hay stacked to retrieve from the day before, I could get a jump on things before the others arrived.

My reward was getting to know the summer sunrise, all pastel blues and pinks. The green, fresh-cut lands were rich with wildlife at this hour. Also, the cool morning agreed with the diesel and with me. I seemed to set the hay down much straighter with nobody else looking over my shoulder.

Several days I got out there an hour and a half before the rest of the crew, which didn't hurt Bill's or Keith's feelings any. On one of those days, Bill showed up at the field about ten and said, "I've got to fly to Great Falls for swather parts today. Want to come?"

"Sure," I said. "I don't have that much in my stomach."

As Bill threaded the 185 through the Belts, he said, "Keep an eye out through here. Always watch for traffic over mountain passes. It bunches up." Sure enough, I spotted a couple of planes, one to the east, going the same way we were, and another, farther east, traveling in the opposite direction. There was no playfulness in Bill on this trip; he knew the price of an error in this country, and he was crisp and precise all the way into the big airport in Great Falls, where the parts were waiting for us, and back again.

Bill deposited me and the parts at the shop. Keith met us there. "Hello, Mr. Fuckoff," he said to me. "There's hay to retrieve, whenever you're in the mood."

The next morning I was out early again, but I wasn't the only one. I was checking the alignment of my first load at the stackyard when the Cessna came roaring out of the sunrise and buzzed the stack low enough to trim a few wayward stalks out of the top bales. When I regained my ability to speak, I got in the cab and radioed Birch Creek One. "Your right tire looks a little low," I said. "Do you always practice your touch-and-gos on top of haystacks?"

Static and a nasty chuckle were the only answer I got, but later I heard from Keith that he'd reported the stacks were dead straight.

"He should know," I answered. "He was close enough to them." And I added more quietly, "Thanks for the help dropping those loads square."

I'm getting to understand Keith's feeling about haying. There's no getting around it, the work is repetitive and it seems endless. We are like leaf-cutter ants, lifting the bales over the backs of the trucks and trundling them off for safekeeping.

By the third week in July we'd moved to Ben Galt's alfalfa, at the part of his ranch known as Battle Creek, named for a battle between settlers and a band of eight Indians suspected of stealing horses on this spot in 1871. We would haul half of the hay home to Birch Creek and stack the other half here for Ben.

As seemed to be the case so often, there was work to do before we could do the main work. We had to clean up the little stackyard where Ben wanted the hay. It hadn't been used for a while, and it was full of old boards, bale strings, and other assorted junk. Not surprisingly, that task fell to Jerry and me.

Then it was back to my old friend the diesel. After working smoothly for the first few days of haying, the PTO knob on the floor was sticking again, and having to engage and disengage it each time

I picked up or set down a load became a real pain. Sometimes I'd have to get out, crawl under the truck and pull on the cable to disengage the PTO.

The field was irrigated by a huge pivot sprinkler system, with pivot wheel tracks separating each land. The tracks were rough on the retrievers. When they were under a load of five tons of hay, the trucks' frames were vulnerable, and jouncing over bumps was dangerous, so I tried to keep crossings to a minimum.

You can't put a vehicle to much harder use than haying. Hot weather, dust, heavy loads, rough terrain and long hours are as tough on machinery as they are on people. Today's weather certainly qualified as hot—the high nineties, not a cloud in the sky—and Willie John and I broiled in the retrievers while Doran and Jerry used the jitney and the Chevy loader to bunch hay for the trucks. We would have finished in pretty good shape, but the Chevy loader threw an alternator belt. Ben didn't happen to have one that fit in his shop, and so it was well after dark when we got the last load stacked, and nearly midnight when I got back to the bunkhouse.

I met Keith at the shop at six this morning, and we headed up onto the bar. It was time to start irrigating again for the second cutting, and this time my job would be a little more demanding. Keith spent a little over an hour showing me the intricate web of ditches and dams at the top of the new seeding. Two deep parallel ditches ran along the south side of the field and supplied the water. As the flooding of each land was completed, the big dams in the feeder ditches were leapfrogged, opening up the water to the next land. It was a big, quite complex field and irrigating it right demanded precision. I also inherited a tiny strip to the south, and the fields on the north side I irrigated the first time around.

Now the days would really stretch out, with many dams to change morning and night and at least half of the hay left to get in. Soon we shifted our haying operation to the Stevens. I was assigned to ferry loads from the field to the stackyard with Lucy. She was rumbling along quite sweetly with her fresh engine, and I enjoyed getting to drive her. One day in the shop I'd given her interior a real spit-shine, first using the air hose to blow out a lot of accumulated debris, then vacuuming up what remained and treating her to an Armor All polish job. Now I was the beneficiary.

She was a little more difficult to drive than the diesel. She steered quite a bit harder and something about her weight distribution made me less comfortable maneuvering her around. The bed had to be loaded with the loader, like the big semitrailers. But the good news was I didn't need to be nearly as precise in the stackyard. I couldn't unload the hay upright as I could with the retrievers, so I just lifted her dump bed and let fly. Keith or Willie would stack the hay with the loader later.

I had only a few loads completed, though, when she refused to start up after Keith had stacked several tons of hay on her bed. After a little tinkering, Keith diagnosed the problem as a bad starter and sent me to town to pick one up from Jeff at Meagher Motor Supply. I always liked going into Meagher Motor. It was just like the small-town parts stores I remembered from my youth, where you sat on a stool just like you did at the diner, and talked to a counterman (or -woman) who knew just about everything there was to know about vehicles and parts. Jeff kept a catalogue of all of Bill's vehicles, and he could produce just about any part you could name on the spot. If a breakdown occurred after business hours, he'd often drive down to the store, open up and get what we needed. Certainly this starter didn't present a problem.

Tyson and Wylie finished swathing the wheel line at the Stevens a few days later, but they had quite a few problems in the huge field with rocks. Swathers, large as they are, are surprisingly fragile machines, and hitting a rock can cause damage. It can also cause the swather head to bounce and miss a chunk of hay.

Therefore, a particularly thankless chore fell to Jerry and me. We took shovels, crowbars and Lucy to the wheel line and picked rocks. Any rock of any size that stuck out of the ground far enough to cause a potential problem was fair game. "I wonder why he's having us do this? It's already been swathed," Jerry grumbled.

"I guess he's thinking about a second cutting," I said. "Or maybe it means he's extending his lease until next year." It was dirty, sweaty, grinding work, tough on the back and legs. As we picked rocks, the balers made their rounds, and every once in a while Bill or Gary Welch would stop their balers and point to a rock nearby, and one of us would hustle over and get it.

We picked four loads of rock. The first two loads we took down to the creek crossing at the edge of the field. The edge of the bank was eroding, so we placed the rocks, one by one, along the edge of the streambed to stop the soil from crumbling away. Fortunately, the last two loads we were able to dump from the bed into a pile near the creek. By the time we finished, about three, Keith had stacked enough hay for us to start retrieving, so we didn't have to worry what to do the rest of the day.

When we started to stack the hay from the wheel line, Lucy overheated and wouldn't start. For some reason the brand-new distributor points and rotor were chewed up. And the Chevy loader had coughed out an enormous plume of white smoke and seized up as Keith was driving it; the engine was blown. The stress and long hours finally caused Keith to blow a gasket as well. He was stacking hay with the Dresser loader when Jerry called him on the radio to ask what to do next. He jumped out of the loader and strode over to where we were standing.

"*Fuck* this," he spat. "Come September I won't be working here. Nothing but bullshit twenty-four hours a day. Change the water. Stack this. Load that. Haul Bill's tractor. Haul Gary's tractor. Make sure you guys are doing your jobs. Keep all these goddamned outfits running. I've had it. You guys get the fuck back to Birch Creek with the retrievers and get them serviced and change your water and stay the hell out of my sight. And you better not have any problems. Don't call me again on that fuckin' radio. I hate answering that son of a bitch when I'm trying to work."

Bill and Gary got the rest of the hay baled at the Stevens, and Jerry and Willie and I got it put up in good order. Keith came back from vacation in a better mood, and even Bill seemed to be a little more relaxed; we were chewing our way through the haying like good little leaf-cutter ants. We moved the operation to Avalanche Ranch next, along Canyon Ferry Reservoir between Townsend and Helena. Avalanche is another of Louise's ranches, managed by Tom O'Donnell.

Tom worked with the Galt boys and my friend Sid Gustafson as a kid on the ranches; he's an old friend of Bill's and was Errol's college roommate. He's respected as an excellent horseman and a hard worker. I'd met Tom about a month before I went to work for Bill, when I assisted Sid, working Errol's bull herd, which winters at Avalanche. I looked forward to seeing him again.

We had a choice of two ways to get to Avalanche: Back south to the nine-mile Y, where the highway to Townsend met Highway 89 to Livingston; or straight up the county road past the bar, past Gypsy Lake, over Duck Creek pass, and down to the Townsend road. It was a lot shorter that way, and a lot rougher, too.

I drove the diesel retriever over that way, with Willie ahead of me in the Chevy. We passed yet another serious fire on Duck Creek. Already crews from several states were working to get around it, and control was expected in a day or so. We made it over the pass in good shape, and we had plenty of hay on the ground waiting for us when we got there.

With the Chevy loader down, we used the jitney and the Dresser to bunch. I got in time on both, then finished the job in the diesel. As we went back over the pass, I realized that I felt comfortable now running any machine on the ranch, with the exception of the Kenworths and the big Cats. That was a big contrast from those winter days when driving anything other than a pickup had my knuckles white.

One morning in mid-August I was shocked to find ice on the hounds' water at seven A.M. A frost isn't so unusual at this altitude, this time of year, but it was certainly a wake-up call. We'd plunged into summer so quickly after the snow melted, just a few weeks ago it seemed, and now we were being served notice that fall was not far off.

We put a new 454 Chevy motor in the red loader that day. In contrast to the way the process had gone with Lucy, this motor slipped in like silk. Having the Chevy back would make things easier.

After Avalanche, we put up the wild hay now the Stevens. We still had wild hay in the beaver slide and the field between it and the corner stack known as the hog lot, where Tyson and Wylie had swathed yesterday; then down by Birch Creek and up on the O'Conner. After that, we'd be on the second cutting.

As we bunched and retrieved and hauled, I thought about coming here with Keith to turn the water on this hay. Since then Keith and Donnie had transformed the hayfields. No more carcasses, just lush green bales.

Up by the Stevens buildings, though, the place was still a mess. Trash covered the hillsides and the corrals—scraps of wood and wire, old car bodies, leaky barrels, feed sacks, baling twine, part of an old squeeze chute, side panels from an old stakebed truck, broken pallets, everything imaginable. Inside the buildings—the shop, the barn, a cabin, a shed, an old train car—it was worse. The train car, festooned with sixties flower-child graffiti, seemed to be filled with almost nothing but old carburetors. Very odd indeed. Apparently for many years every time a carb got changed out, the old one had gone into the train car. The shed and cabin were more conventionally dirty and cluttered. Just walking by the buildings gave me a prickly feeling. I knew how the mess must have made neat freak Bill Galt cringe, and therefore I also knew it wouldn't be too long before I was shoveling it all out.

I didn't have long to wait. The entire crew got the assignment, and it took all of us three days. Keith moved the car bodies and a few other large items with the loader. Then he, Willie, Jerry and I spent half a day cleaning out the shop. It was like a huge scavenger hunt, looking for usable items and throwing the rest away. Keith put the bucket in the doorway to fill with trash, and soon we had three big piles going in the yard: one to burn; one with nonflammable trash, which Bill had contracted to be hauled away to a landfill; and one heap of salvageable scrap metal. Every square foot of the floor needed scraping to clean up caked-on dirt.

The barn and the corrals were probably the worst of it, just dirty, heavy, mindless work, collecting a decade of somebody else's neglect and rot and putting it into ugly piles on somebody else's say-so. As we carried the truck panels into the scrap-metal stack, I dropped one end on my right foot, and for the rest of the day my boot was spongy with blood and two of my toes felt like they were going to pop through the leather. I figured they were broken and I also figured I'd better not take my boot off because I wouldn't be able to get it back on.

When I got home that night, the toes were indeed purple and enormous and sore as hell. The next morning, I bandaged them and jammed them into Nikes; boots were out of the question. The last cleanup day was spent on the cabin, the shed and the hillside. By six-thirty, the two trash piles were fifteen feet high by twenty feet square, but the hill looked like a hill again. The barn and shop and cabin and shed were usable, and suddenly, looking around, I felt my spirits climb. There was something about ennobling this place, restoring its dignity, that transcended the plod and grunt of our labor.

The rest of the haying went reasonably well. The diesel retriever, which had performed so nobly for six weeks, finally broke down. One night at Catlin on Ben Galt's ranch I picked up the very last load in the twilight and suddenly the engine began running a little rough; then the oil-pressure gauge swung to zero.

I grabbed the key and shut it off, then called Willie on the radio and told him what had happened. "Turn it off!" he yelled, and I assured him I had.

When we investigated, we were hugely relieved to find a vacuum line had popped loose, and I regained oil pressure immediately when it was reinstalled and the truck was jump-started. But it was still not running quite right; the vacuum pressure would not hold when my foot was off the gas, and with the batteries weak I knew I probably couldn't restart it if it died, which made an adventure out of the drive back to Birch Creek in the dark, sans headlights.

We were ready for our last baling job, the second cutting up on the bar, just before Labor Day. I had to fix a flat on the Chevy retriever one morning, which was entertaining, since the wheel was the old split-rim variety and the tire was a tube type. It was a bitch to break down and even harder to reassemble, but I'd just about finished it when Bill Galt came in and found me kneeling by it, holding the air nozzle on while it inflated.

"You know, Mr. Goodwrench, when I was in high school, a hired man at Dad's ranch at Utica had his head in front of a split-rim wheel while he aired it up, just like that, and I was given the job of cleaning his brains off the wall," he said. I got up hastily.

"Don't ever do that again. Those things are incredibly dangerous. Make sure you don't have any part of your body in front of that wheel when you do that, and point it away from other people." So many ways to die on a ranch, I thought. I'd tried a few of them, and more by good luck than anything hadn't succeeded yet. Finally Bill and Gary baled the last hay of the year, and Jerry and I started ferrying it back to the nearly full corner stackyard. I nursed the diesel through; weighing loads was especially difficult, but I managed to keep the engine running. "I'll fix it when we're through," Keith growled. "Just don't let that son of a bitch die."

Jerry went with Keith and Willie to move the swathers up to a field where Keith and Willie were going to cut some straw, leaving me to retrieve bales into the evening. Because I was running at the slow speeds necessary for safety in the retriever, each load was taking me about fifteen minutes, and at seven-thirty I took a look at the light in the west, still glowing purple and blue on the horizon, and figured I could get three more loads if I was lucky. As I took the first one into the yard, Bill passed me, headed for home. "You'd better knock off," he radioed.

"I think I can squeeze a couple more in, Bill," I replied, to which he responded, "Whatever blows your hair back," but I could tell he was pleased. The last load was cutting it close, and I nursed the diesel back to the shop in the dark, greased and fueled it, and knew I'd finished my last hard day of haying.

Somewhere near six thousand tons of hay. Two thousand six hundred *miles* of baling twine around fifteen thousand bales. Assorted fan belts, water pumps, fuel pumps, starters, engine blocks and tempers heated past the breaking point. We were done haying. All but a few bales of straw, which in the case of Bill and Willie John proved to be the straw that broke the camel's back.

On Saturday, Willie and Keith had gotten the straw up to the stackyard, but it needed to be stacked. Willie had said he'd do it on Sunday.

On Monday morning, Bill met Willie on the road and asked him to move a couple of cows. The conversation that followed went something like this:

"I still need to stack that straw."

"I thought you were going to do that on Sunday."

"I decided I'd rather spend Sunday with my family like everybody else did."

"That's fine, but don't say you're going to do something and then not do it."

"What difference does it make?"

"None to you, I guess, because I won't need you around here anymore. I need hands I can trust."

"That's right. I quit."

"You sure do."

Nobody was all that surprised. Willie and Erin weren't happy with the hours. Bill wasn't happy with Willie's attitude and with several incidents where he felt Willie had been less than forthright with him. Still, it left everybody a little disturbed. Jerry sidled up to me in the shop and said, "You hear about Willie?" He looked scared, and he was honest enough to say, "I figured I'd be the one to get it before him," and I liked him for that.

Jerry was perplexing. I figured it couldn't be much fun being the one who got yelled at the most. I'd advised him to speed it up, try to show some hustle, get on Keith's good side, but it just seemed to get worse. He'd go along okay for a day or two, then be late to work or forget to do something he was told or pull some crazy stunt, and he'd be back in the shithouse with Keith.

That morning Keith shook his head. "That fuckin' Willie," he said. Then he turned to Jerry and me and said darkly, "You'll find out Bill will stand for only so much shit and then it's down the road."

The news seemed to put Keith in a foul temper, and I could understand why. Not only did it mean more work for him, but he had been able to talk to Willie on a level beyond what he could share with anybody else on the ranch. Keith and Donnie didn't get along, and Donnie wasn't around Birch Creek much anyway. And while Keith and I were generally pretty friendly, and I really liked and admired him, I knew there was a lot we could never share. He and Willie were both White Sulphur natives, knew a lot of the same people, had shared many of the same experiences.

They were alike in many ways and very different in others. For Keith, the most important thing was doing the job right and being utterly dependable and responsible. Willie played a bit more of a lone hand, despite his talent at the work, and now it had come back to bite all of us.

I would miss Willie sorely. He'd taught me quite a bit, and we'd worked side by side for a long time. He'd fished me out of the creek when I tipped the four-wheeler over, showed me how to pull a calf, how to run the squeeze chute, how to improve my wire splices, how to use a torque wrench, how to use starting fluid to pop a bead on a tire (don't try this at home), how to hook up a fifth-wheel trailer, how to notch a fence rail, how to throw a Houlihan loop. Sometimes he'd snarled at me, usually for good reason. The rest of the time he was patient to a fault, and I appreciated his rather cynical sense of humor. Where we'd found areas of common interest—fishing, baseball, pool, horses, high school football reminiscences—he was engaged, even effusive, in our conversations. Willie was intelligent, articulate and, much of the time, pretty angry.

By the time I saw him later in the day, he was philosophical. "You know I've been looking around for another spot anyway," he said. "I've got a couple of interviews. And it'll be nice to have some time with Erin, Beau and the baby."

Maybe it was all for the best. Bill spoke frankly to me in private. "I like Willie," he said. "It just wasn't working out; he was taking advantage of me. You know when that baby was born I told him to take as long as he needed, but to keep in touch with me. I didn't hear a word from him for seven days. Just a phone call would have been nice." He sighed. "It's too bad, but I think it was time for Willie to go. He'll be happier and so will I. And I wouldn't be surprised if he came back sometime."

"So you'd hire him back?"

"Sure I would, if I think he's had some time to figure out what went wrong. Willie's pretty handy. Not as handy as he thinks he is, maybe, but he's a damn good hand."

This had been Willlie's second tour of duty at Birch Creek. Looked at from that perspective, his departure seemed like part of the rhythm of the place, like Keith's and Donnie's occasional leave-takings. It's no wonder these men got sick of each other from time to time, I thought, sealed together in this never-ending work cycle like pickles in a mason jar—a hundred-thousand-acre mason jar.

JULY 14

Bill took pity on us today, or rather the weather did; a rainstorm left the cut hay too wet to bale, so we got this Sunday afternoon off. We'd been haying on the Gurwell, and seeing the tree line of the Smith River had been driving me crazy. The fields were singing with grasshoppers, and I knew that the brown trout had to be feasting with the gusto of Paul Prudhomme eating popcorn shrimp. So about seven o'clock I put on cut-off Levi's, old tennis shoes and a T-shirt, tossed my rod in the GMC and drove down to the stackyard on the river side of the Gurwell. It was a short stroll down the bluff and through Elmer Hansen's pasture to the river.

The mosquitoes were murderous, but I didn't care. The thrum of hoppers was constant, and every twenty seconds or so I would hear fish-splash noises that ranged from delicate to the sound of a bowling ball being dropped into a bathtub.

I am not a particularly good fly-fisherman, and I did not need to be. The browns were big and jumped wild as spurred broncs, and I caught more of them than anyone needs to on a summer evening. I fished until the deepest part of dusk was turning to night, keeping only the last—two pounds of dinner with a yellow belly and wide dark spots across its back. I fried it with butter and garlic and green onions and fat slices of an enormous meadow mushroom I found on the way back to the truck, bigger than the trout and just as firm and pink. I drank two Miller beers and wished for this one night to come back to me one more time, perhaps late in my fifth decade, when I would need it just as much as I had tonight.

Part Four

the best of all seasons

chapter 19 • more fencing, more progress

IN THE SPRING, WHEN THE TRUDGE OF FEEDING AND CALVING heifers gave way to branding and irrigating, it was possible to perceive the beginnings of a cycle, a tuning of work to weather to biology, thawing, nursing, greening, estrus, breeding, sunshine, water, growing. Now another cycle was beginning.

The cooler air was refreshing, and despite my broken toes I felt as good as I ever had. I realized with a real jolt of pleasure that what I felt was the arrival of fall, by far my favorite season here in the Northern Rockies. It is often brutally foreshortened, but it is spectacular while it lasts.

Fall also brought a sense of survival and accomplishment similar to what I felt at the onset of spring. It was a sense of growing awareness of my surroundings, of what we are about and what must be done. I can't remember learning so much about anything in such a short time. I was much more confident day to day. I felt—dare I say it?—downright handy.

Keith gave me an appropriately transitional task. "David, take my truck and go pick up the dams. Start at the field next to Bill's house, then the one across the road, then the Gurwell, then up on the bar. And don't take the truck out in the alfalfa."

It was muddy work, after this morning's rain, and it felt like there was more to come for sure. The wind was kicking a little and the sky was the color of wet slate, but the temperature was perfect for strenuous work like this, about sixty-five degrees, and despite the throb of my injured foot, this strange euphoria persisted. I felt as though I had come out the other side of something. Perhaps it was the neatness of it, the little completion, putting the dams out in the fields with Tyson in May, now retrieving them, bundling the sticks inside for carrying, throwing the bulky rolls up across my shoulders and humping them back to the pickup, then moving to the next ditch and the next and the next. Sometimes the dams would be at the near end of the ditch, sometimes at the other, sometimes in the middle, so it required a fair amount of walking, peering down ditches for a glimpse of orange, followed by the same bundling, throwing and humping.

It was gustier up on the Gurwell, wild and rugged today as the coast of Cornwall, a place and time where it was possible to be raucously alone, to sing Hank Williams songs loudly and badly, knowing the clamor would be snatched by the wind six words at a time, and heard nowhere: "Hey, good lookin' . . . whatcha got cookin'?"

We have less than a month before we need to be weaning. The sign was everywhere: whatever you want to get done before winter had better be tended to damned soon. I'm sure ranching in South Texas and Arizona has its own special challenges, but one of them is not looking over your shoulder for snowflakes in September. Up here, so much has to be done in such a short time.

Right now, the biggest thing was fencing. We needed to move a lot of cattle around to various pastures in the next few weeks, and they'd all winter around Birch Creek, too. You didn't want to be out fixing fence and trying to find wayward cattle at risk in the snow and cold, so fences had to be fixed now.

One morning when Keith and I were both at the shop early, he said, "Let's take a ride," and we headed onto the bar. Keith showed me the big pasture that ranged behind the calving shed, past an old homestead on George Zieg's spread, down in the hollow beneath Kite Butte, and up here onto the bar. The pasture zigzagged up and over a ridgeline, across the creek and next to the back of the heifer pasture. Keith showed me where to start and said, "Just don't go out of this field. Get all the way around and you'll be fine." He reiterated Bill's fencing philosophy: Take as long as you need to get around a fence, but it better hold cows when you're done. If it doesn't you'll be out rounding up cattle and fixing the fence on your own time. He added, "We've got a lot of miles of fence to get around. You and Jerry will be fencing for the next month."

The pasture Keith showed me took all day. I had to go back to Birch Creek for more steels once. I picked up five more bundles, and before we were done all but three of those posts were in the ground.

"When you get to the headgate, give me a call on the radio," Keith had told me the next afternoon when we started out to fence. "It should take you a couple of hours." That was three and a half hours earlier, right after lunch, and Jerry and I hadn't seen a headgate yet. What we had seen was a lot of shitty fence running along a creek bottom and therefore susceptible to washouts, overgrowth and deadfall, as well as banging around by various critters that traversed the brush. It was impossible to drive this fence or use a four-wheeler; the brush was impassable for long stretches. The only way was to bushwhack through on foot and that's what we did.

Jackfence, or jackleg fence, is good for really rocky or swampy ground because you don't have to set posts. It's created out of crossed timbers, with wires strung on both sides. A lot of this fence was made that way, but the wires were long crashed down. Most of it just took new staples, but a few of the timbers needed replacing, so I retraced my path through the brush and got the chain saw from the truck.

In other spots steel posts had been set into the edge of the creek and had been bent over almost flat by something, probably moose. Some I managed to straighten; most had to be replaced.

When we got up onto the flat and away from the brush and the wet ground for a while, I thought the fence would improve, but apparently this spot was much used by wildlife, and the top three wires were down for a fifty-yard stretch.

I spliced and stretched all of them. It took about fifteen minutes, and about as long again to make it to the headgate and more bad news. I called Keith, who said, "How does it look?"

"Not good," I replied. "It's jammed with sticks and debris."

"That's what I figured," Keith said. "It's got to be leaking. You need to clear it and get it shut down."

The water was numbing cold, and I was in it up to my shoulders for half an hour, pounding at the obstructions with a sharp rock, trying to clear it and finally succeeding. "Well, no shit," I said wearily as the headgate slipped home. "Let's get back to the truck and get out of here."

But we hadn't had enough fun yet. As we started to walk back, we heard something heavy rustling disconcertingly close in the brush behind us, followed by a bugle so loud I thought at first it had to be Jerry, with an elk call. I looked up just in time to see six elk, led by two big bulls, crash through the fence I'd just repaired.

I couldn't believe what I'd just witnessed, maybe fifteen yards away. One of the bulls was the biggest I'd ever seen, with an enormous wide branching rack, six points on each side, so big it seemed impossible for him to carry. It gave him a look of majesty out of all proportion to the everyday world, like Secretariat prancing around with the Kentucky Derby roses. Awestruck, I watched him for the fifteen seconds it took him to disappear into the trees. I swear he turned his head sideways to fit between two lodgepole pines just before I lost sight of him.

It was a few moments before I could turn my attention back to the fence. When I did, I discovered a wooden post was broken and two wires were down. "Not our day," Jerry said, and I nodded, but my heart wasn't in it. I'd give another half hour of work any day to see what I'd just seen.

A fence stretcher is a wonderful thing. I'd done a little checking, and found out that the modern version was invented by farmer Bill Greutman of Miller, Nebraska, some time between 1910 and 1920. Still manufactured by a Nebraska firm, the Dutton-Lainson Company, it is deceptively simple and primitive-looking. It is very well thought out indeed, and there is nothing better to splice or tighten barbed wire with.

Two serrated grips attach to the fence, about a foot and a half apart. Then a lever is used to ratchet the clamps closer and closer together, putting tension in the fence. Then the two broken pieces can be spliced, with another section inserted if necessary, or the one loose strand can be cut and spliced tight together. If need be, one of the stretcher grips can be attached to the side of a brace pole or corner post or gatepost to tighten wires against.

When the wire is tight and the repair is made, the ratchet is released and the stretcher removed. If you've done the job right, the fence is tight as a fiddle-string.

As we fenced our way around one Birch Creek pasture after another, it occurred to me that I was getting better at this, too. It was taking me less than half as long as it had in the spring to make repairs, and they were better done. I still found fixing fence to be one of the most satisfying jobs on the ranch. I was getting a better and better idea of the warp and woof of this place, of the way the pastures followed the shape of the land; which fences were section divisions and which were placed for different reasons relating to water and rock and slope.

It was a good primer for moving cattle; now I knew what Bill meant when he said, "I flew over Section 19 and saw two cows that shouldn't be in there. Move them over into 30." Knowing the fences and corners, knowing where the gates were, could save a lot of time and trouble.

Tonight, when I got back to the shop, I met the newest member of the crew. He is a friend of Jerry's, Brent Wanderi, a husky Norwegian from Minnesota who came out this way doing construction in the summer. He'd met a girl from White Sulphur and decided to stay around for a while. He was servicing one of the pickups, and it didn't take long to see he knew what he was doing. I figured he must be about twenty. He is friendly but quiet, and has an air of efficiency about him. He'll be a decent hand, but no match for Willie John Bernhardt.

We'd been pounding steel posts around this ranch at an amazing pace. I'd gone into Jack's Ranch Supply twice in the past week to pick up truckloads, and finally we'd run them out. "Are you replacing good posts?" Keith demanded. "Don't replace a post unless it's pretty bad."

"I'm only replacing them if they're waving in the wind, rotted off at ground level," I said, and he nodded.

"That's right. I guess it's just the year to pound posts."

The next morning I went into Mile High Hardware—irreverently known as Sky High—and got all they had, seventy-five posts. Brent, it turns out, grew up on a dairy farm and was no stranger to hard work. He'd done quite a bit of fencing, too, that was obvious, as we went around the lower meadows of the O'Conner together, from the stackyard a couple of miles up a long, steep hill. The day was warm and sunny, and the wind blew strong all afternoon, in grass the color of honey. A few cows ventured down from higher ground to investigate us, then munched their way back up into the timber. They would have to be taken out of here soon, I thought; otherwise the snow would make the gathering very tough in this terrain. The pounder I had was too light; Brent had cagily snagged the one I usually used. That meant a lot of ergs expended to achieve the desired result, eight bundles of new posts in the old ground. Forty more places over three miles of fence that would hold cattle now and wouldn't yesterday.

Bill came up to me the next afternoon. "The Cessna just had its annual inspection. The first trip after the annual is the one that kills you. Let's go flying."

The plane seemed to want to fulfill Bill's morbid prophecy. As we left the ground the nose suddenly plunged downward. Bill fought it back up with a shake of his head. "Something not right with the trim tab. I'll get it checked out."

The rest of the flight was terrific—a sunny, breezy fall day. But

I soon found out I wasn't just up here to look at the scenery. "Where are we?" Bill demanded after a few minutes.

"Let's see, that's the highway, that must be Newlan Creek Road. We're coming up on the River Field."

"Right. Now I want you and Brent and Jerry to go around that fence. Be careful going between those lands. There's a lot of swamp in there and I don't want anything stuck, do you understand?"

"Yes."

He flew low over the huge field and I could see the crossings clearly, higher, drier spots in between the lands. "See that?" Bill said, pointing to a light-colored strip of fill that ran across the low marsh between the last two lands. "That's the only way across, so don't try anywhere else."

"Okay."

We flew for another few minutes. "Now where are we?"

"That's the corrals there, and the steer pasture and the fat pen."

"Okay, what do you see below now?"

"Looks like a gate down at the back of the fat pen."

"There's two of them. One there near the corrals and—"

"—the other one to the northeast of the calving barn there."

"Okay, I want those fixed and then you can start on the river field."

The gates were quick. The River Field was another matter. We started at the northwest corner. Before we split up I said, "Be careful about crossing lands. Both of these first ones have crossings close to the center of the field." Jerry took a four-wheeler and went south, I walked east along the road, and Brent took the truck and drove over toward the east fence.

The leg I started with was in good shape, and about eleven o'clock I hit the corner and caught up with Brent, who'd already put in fifteen posts. We started working southward, presumably parallel and across the field from Jerry. This fence was very bad. Apparently every other post had been replaced in the last few years, but many of the old ones were rotten. So we inched along, leapfrogging each other as we pounded, stretched and clipped.

Jerry came over to our side at noon, and we ate lunch together.

His four-wheeler looked like it had the afternoon Willie and I had bushwhacked through Section 19. He'd missed one of the crossings and stuck the four-wheeler thoroughly, but had managed finally to horse it out. "Don't tell Bill," he said. "I'll wash this thing after work tonight." The wind came up in the afternoon, and by three it was gray and cold. The field was a long rectangle, and the two north-south fences were the long sides and in the worst shape, particularly on the low edge of the lands and in the areas between, which even after the dry season were seas of mud.

The wind was constant, draining us of energy like a sickness, and I figured we were no more than halfway around when it got too dark to continue. We'd run out of steel a couple of hours before, so we'd marked bad posts with plastic ribbon; we had at least twenty to put in before we could push ahead in the morning.

October arrived on the west wind, which had not lessened; the bunkhouse had groaned and whined all night like an old hound having a bad dream. Finally I gave up around five and had some breakfast.

As I fed the hounds, sleet whipped into the windward side of my face, stinging like the guilt of tiny sins. I gassed the truck and the four-wheelers, reloaded with twenty bundles of steel posts—Keith had taken the flatbed into Jack's and picked up an entire pallet the other day—and drank an extra cup of coffee as I waited for Jerry and Brent.

It was a dirty, miserable day. As we came up out of the coulee and onto the bench at the back of the river field, the wind doubled in force. I was stretching wires on a gate in the corner, reefing on the second wire, when the gatepost broke under the strain, and as I fell backward, the wire whipped me across the face, taking a chunk out of my chin. It was little worse than a bad shaving cut, but it bled for the rest of the day. I had a post on the back of the truck that would work, and managed to get it set in right and the gate rebuilt.

We finished around six, and I was damned happy to see the last of this particular fencing job.

* * *

Bill had found something else for me to clean: the chute shed, which was desperately in need of it, to be sure. Jerry and I gave it the same treatment we had given the calving shed six months ago, the Stevens six weeks ago, the barn and various other spots. Instead of a slash double O, I thought, perhaps Bill should have a broom and dustpan for a brand.

It was a relief, then, to get a fencing assignment: a single section of land on the way to Jackson Lake and the O'Conner. The land belonged to George Zieg, Bill's neighbor, and Bill leased it from him.

An old friend of Bill's, Mike Miller, needed a place to winter forty head, and Keith had trucked them to Birch Creek yesterday from Miller's place near St. Ignatius, Montana. After we vaccinated and preg-tested them, they'd be parked in this pasture, called the Lone Section.

It was a rough terrain, wild and unpredictable, with deep coulees and rocky outcrops. Much of the west side of Kite Butte was within the section, and the other end of the section, toward the O'Conner, overlooked Jackson Lake.

This was fencing graduate school. The first two legs of the fence had an insatiable appetite for steel posts. The high ground was incredibly rocky, and the fence was sagging and weatherbeaten, with posts and wires on the ground for long stretches.

The third leg was mostly jackfence, over boulders, and fortunately it was in pretty good shape. I was feeling pretty good about things until I got to the corner, which dropped away into a ravine so steep I had to park the four-wheeler and pack posts down by hand. It was nearly six P.M. when I finished the corner, hiked back up to the four-wheeler and started on the final length. I'd saved the leg that went across Kite Butte for last. I'd looked at this butte almost every day I'd spent at Birch Creek, and in the last month, fencing, from almost every side, and I'd always wanted an excuse to see it up close.

I'd been curious about the name, and I'd asked a few questions. George Zieg, born practically in the shadow of the butte, said he'd always heard it called Walwark Butte. Checking with Meagher County's foremost historian, Lee Rostad, I discovered that George

Walwark had been a schoolteacher at Thompson Gulch in the 1870s and had probably lived near the butte. But rancher Ron Burns told me that Zieg's father, George Sr., had told him this story of Kite Butte:

The flume, or water-carrying ditch, for the hydraulic mining operation at Thompson Gulch had been built by Chinese laborers, as had so many daunting projects in the West, including many of the railroads. According to some accounts, one thousand Chinese had been paid a dollar each per day to build the flume, and the job was done in two days. The flume remains essentially intact today.

Apparently some of the miners had built some sort of a primitive glider or kite, and they wanted to try it out, and supposedly they chose one of the Chinese laborers as the subject of their experiment, strapped him into the contraption and tossed him off the top of the butte.

Burns said Mr. Zieg senior had no idea how well it had worked, or of the fate of this early aviator. As I wrestled the four-wheeler along the last fenceline of the Lone Section, I sympathized with the poor flier. Lava rock jutted up through the grass every few feet; this would not be a hospitable landing area.

Fortunately for me, this was the one side of the fence that had been worked on recently. Someone had replaced this entire fence within the past year or two; I didn't find as much as a clip loose all the way back down the slope of the butte to the gate where I'd started.

Keith took Brent, Jerry and me fencing the next morning, up past Jackson Lake and onto the O'Conner. The grass was white with frost, and though I wore coveralls and a warm coat, the ride up the hill, into the wind, was chilling.

We met at the gate at the back of the Lone Section. Keith lined us out there, making sure we were completely outfitted and clear on where we were going before he headed down to the corrals. "Go up to the next gate and get around that next pasture," he said. He showed me how to tie a bundle of wire onto the back of the four-wheeler to keep it out of my way. "Whose radio is this?" he said. Someone had left a handheld in his truck.

"Mine," Jerry said, and Keith handed it to him, and said, "Everybody clear on where to go?"

It warmed up a little by late morning, but the wind kept us from becoming too comfortable. The fence was better than the Lone Section, and by shortly after five we were finished.

On the way back, I stopped at the gate to the Lone Section. I had three steel posts left, and I'd been pretty thrifty with them when I'd gone around this fence yesterday. I knew just where I could place them to good effect, so I got off the four-wheeler to do that.

Jerry, in front of me on the way back, had been cruising much slower than usual. Now he came up to me when I stopped to offload posts. "David, come on back to the shop. It's quitting time, and if you're back, we'll probably get to go home."

"Sorry," I said. "I'm here, these posts are here, and I'm going to pound them. I'll meet you back there in a few minutes."

Contrary to Jerry's plans, by the time I returned to the shop, Keith had given him a couple of tires to fix. Keith's pickup was in the shop, awaiting my attention; it needed an oil change and service. Brent, meanwhile, was fixing a shock absorber mount on the crew cab. It was nearly dark when we were finished. I took the handheld out of my coveralls and put it in the charger, and I noticed that one of the chargers was empty. "Where's the other handheld?" I said. "Brent, did you have it?"

"Nope."

"Jerry?"

"Yeah, I did, and I don't know where it is," Jerry said unhappily.

"What?"

"Keith handed it to me up there by the lake, and a few hours later I realized I didn't have it with me. I looked all over for it up there. I don't think I ever had it on me. I think I set it on the back of my four-wheeler and forgot it."

"You mean it's up there lost somewhere?"

"Yeah."

"Jesus, it's almost dark. I'm going up there right now."

"I'll look for it tomorrow. I've got to get into town tonight."

"Jerry, if the battery wears down overnight we'll never find it. Did you tell anybody it was missing?"

"I was going to call Bill later."

"Those things cost eight hundred a pop." I called Bill and let him know where I was going. Anytime the crew was out late, he wanted to know about it. "Jesus, yes, get up there," he said. "Take Jerry and Brent with you. One of you can key a handheld while the others walk and listen."

We raced up there in the maroon truck. I remembered Jerry zipping around me this morning as we left the Lone Section, wheeling off the road and through the sagebrush, and I couldn't see how the radio could have stayed on the four-wheeler through that, so we started right there. I figured we were literally looking for Jerry's job. Bill and Keith would be furious, I knew.

"We ain't gonna find it," Brent said rather mulishly. He was irritated at having his day extended. I wasn't too pleased, either, but I knew that any chance we had of finding it would be much reduced by morning.

We walked through the dark with flashlights, speaking into the handheld. "One, two, three, four, five," I intoned, listening for feedback from the ground in front of me. I knew that Bill was listening to every transmission, along with everybody who had a radio: Keith, Donnie, Pat, Julia, Bill Loney, on and on. Normally Bill frowned on radio traffic this late at night, but obviously this was an exception.

Brent's pessimism proved to be justified. We looked until after ten P.M., to no success, and finally Bill told us to knock off.

I called him on the telephone when I got back to the bunkhouse and told him that on the way back, Jerry had said he didn't even think he'd turned the radio on, which made our efforts tonight all but pointless. No wonder we hadn't found it.

Jerry was curtly assigned radio-hunting duty the next morning, while Brent and I were sent to fence the big pasture known as Little Birch. Probably eight miles of fence around the perimeter, plus a cross fence bisecting it. "Leave the gate by the cattle guard open," Bill said. "We'll be parking cows from the O'Conner in there."

In this huge field with creek and brush meandering through, it struck me forcefully how much the land had changed in the last few weeks. The caramel color of the range grasses was the same, but everything else had changed dramatically. The lush green of cottonwood and willow by the creeks had turned to orange and red, the sky from blue to deep gray, the water from blue to black in the dull light. To complete the transformation, wisps of silvered ground fog lay in the creek bottoms this morning, damp and cold as I passed through on the four-wheeler. It was as though I were visiting another, more mysterious country, and I knew that sometime within the next month, all I could see now would be white.

Brent went one way, I went the other, and I didn't even catch sight of him again until mid-afternoon. I'd gotten from the front gate west to the corner, down past the cross fence, then within sight of the back of the field, but I needed more posts. I radioed Brent. He was out, too, so we made the forty-minute round trip to the shop yard for posts and back, finished the perimeter in a couple of hours and then tackled the cross fence. Wylie and Willie had built it only a couple of years ago, and so it didn't take too much.

When we got back to the shop, Keith said, "Let's all go look for that radio." And so we spent from six P.M. until dark, again fruitlessly. "We're off tomorrow," Keith said, "but if I were you, Jerry, I'd be out here looking."

Brent and I got the chore of fencing around the alfalfa fields on the north side, and the neighboring wheat field. Brent set off around the wheat, and I worked the alfalfa along the county road. The fence was not awful, but irritating. Nothing major was wrong, but I couldn't make very good time because every other post needed one or two staples. And at the end of each of these ditches I knew so well, where the water flowed steadily for days during irrigating, I could count on finding one or two rotten posts.

Bill buzzed us in the Cessna and radioed us about three cows he saw in the other side of the field we were fencing. He asked us to go get them when we were finished fencing, and bring them in to the corrals.

This field stretches clear over to the Dry Gulch, where it corners and turns westward. The fence was much worse on the leg that turned away from the road, and soon I was wishing for that irritating alfalfa fence again.

I cleared one rise in time to see Brent negotiating a hill so steep I probably wouldn't try it at all. He was doing it in reverse gear. I caught up to him a little later and asked him about it. "Something snapped in my rear differential," he said. "That's the only way I could get up the hill." He flashed a smile. This kid could drive; he spent a couple of seasons on the snowmobile-racing circuit, so four-wheelers came naturally to him.

I found myself warming to him. He was an extraordinarily hard worker, and although he was friendly he maintained a certain almost dignified reserve beyond his years.

I had fenced the Dry Gulch before, and I knew we were about to encounter some trouble, in the form of a swamp. The ground was marshy enough to make the fence unstable, and it got a lot of traffic from animals coming to crop the grass, which was always green because of an underground spring, and to drink when water pooled on the surface.

We finished soon enough, and went back after the cows, which we found in the wheat and hazed over toward the alfalfa stubble. We waltzed them through the various ditches and out onto the county road without a false move, Brent working the high side while I pushed them along the fence.

Bill met us at the corner by the bunkhouse and asked me to finish getting them into the field in front of the corrals while Brent got a few things done in the shop. His mechanical ability is an increasing asset.

As soon as I got them past Keith's driveway, I zipped by and opened the gate into the corral, then got back behind them and walked them up. The fat little dears went in without a murmur.

First early winter, then Indian summer. The gods of fall were smiling, and we got a couple of days that were 70 degrees, windless and sunny. Pretty nice weather for fixing fence or chasing cows or just about anything, for that matter. Like floating the Yellowstone in a drift boat, throwing streamers at big fall-spawning brown trout, for instance. Oh, well.

Instead, I fenced stackyards around Birch Creek in the morning and headed to the Stevens after lunch. I went past the now-familiar osprey nest, down the road Willie and I had traveled the day I stuck the truck, and went around the pivot field, fixing fence and putting out salt. I finished in about an hour and a half; the worst spot was a place where the fence had been cut to get a combine in. I knocked in a couple of posts and strung four twenty-foot lengths of wire, and it was a fence again.

I ran up the Clear Range road and met Donnie. "Come on," he said. "We've got to fence that little field for Ginger's cows." I knew the field he meant, a little corner between the wheel-line field and the lower hayfields. "You take this leg down through the creek and I'll make it around the rest," he said.

"Okay," I said. "I'll come bail you out after I'm done."

He just laughed, and pretty soon I saw why. The leg of fence he'd given me ran through heavy brush down to the creek—or at least it was supposed to. It looked as though it had been knocked down for years. I put in every post he had in the back of his truck, which was just enough to get me to the creek crossing. There was nothing left of that, either.

I remembered fencing across Little Birch with Willie and Tyson. I stretched wires across, found an old fence rail and dropped that from the wires, got around the rest and had the pleasure of seeing Donnie's jaw drop when I came walking out to help him finish his part.

"Did you do the creek crossing?" he demanded.

"How do you think I got wet up to my ass?" I said. "It's done."

"Well, hell. Finish pulling up this hogwire along here and I'll go get those cows and run 'em in," he said.

"You want a hand?"

"Nope, just finish this up. I'll be back with them in a few minutes."

Just as I finished, I heard the sounds of cattle moving, the excited, protesting low grunts they make, the shuffle and thud of many hooves, and Donnie's voice soaring over all, staccato: "Hi! Hi! Hi! Get up there!"

I turned and watched these perfectly muscled Angus beauties streaming into the little field, outlined against the sunset, orange light spilling around their black backs and heads like God's own cows, the dust from their running softening the scene into something classical and painterly, a real goddamned Kodak moment if ever there were one. Then I knew I'd rather be doing exactly what I'd done today, creek and brush and heaving salt included, than anything else I could think of—even floating down the Yellowstone in somebody else's boat like a sporting john from some coast or other.

So much for Indian summer. After a couple of days the west wind I'd learned to dread so much last winter began to blow, and by eight o'clock it had brought a hard, cold rain, which tethered us to the shop. By three P.M., Keith was running out of things for me to do. I knew this because he had me sorting hex bolts in the pigeonholes on the south wall to make sure, for example, that there weren't any 3-inch-long half-inch bolts hiding in the compartment with the 2½-inch-long half-inch bolts. Agh. Jerry, meanwhile, was going through cans of old bolts and washers, trying to salvage anything useful. I was fighting to keep a grip on my sanity when the phone rang. It was Bill. After they talked for a while, Keith hung up and said, "Okay, you guys, get the white Ford fired up and loaded with fencing gear."

Fired up. Yes. The white truck always demanded a jigger of gas straight to the carb, or at least a shot of starting fluid and, of course, jumper cables. This time was no exception, but when Jerry doused the carb and I cranked it, we got a backfire and a bloom of flame the size of the truck. Yeeow. Jerry got some singed eyebrows and a spike in his heart rate that would have brought cardiac-ward nurses on the run.

Our assignment was the fence around the bull pasture and on up the county road. As we left the yard at three-thirty, Jerry told me he had to be back at five because he had an errand to do in town. "I've already cleared it with Bill and Keith," he said.

So I gave Jerry a ride back down to the shop at five, picked up three more bundles of steel posts and headed back out. I fenced along the highway for the next two hours, and they were the best of the day. Maybe it was partly from being cooped up in the shop, maybe partly the pleasure of being alone with the country once again. Whatever the context, this was a feast for the senses: the smell of the land after rain, full, earthy, constant, with wafts of sage, fresh-cut alfalfa and somewhere upwind the faint musk of an animal I couldn't identify; the welcome pliancy of the moist soil as I pounded posts; the little nick of wind against my face as I bent to clip wires; the softening light of evening on the bouldered expanse of the Big Belts; and best of all, from those deep shadowed defiles, a Wagnerian opera of howling coyotes and bugling bull elk.

When it got too dark to fence I climbed into the old white truck, turned the key to zero effect and congratulated myself on having the foresight to park the goddamned thing on a hill steep enough that I could coast down, pop the clutch and avert a long dark walk to the shop.

SEPTEMBER 30

The meat freezers are getting pretty sparse, so a couple of the boys on Death Row are going down today. I went into the fat pen and fed them a bale of hay, up by the road, a little guiltily, to lure them to an area where Doug Pierce, a meat cutter from White Sulphur, could slaughter and get them out of the pen easily.

The hay ploy didn't work; there was still enough grass in the pasture that the fats weren't all that interested. Also, I'm sure that seeing one or two of their number killed and gutted every three months or so had made them a little suspicious.

Pierce and Bill Galt had taken up shooting positions. Bill let loose first, shooting twice at one steer, then twice more at another. Pierce administered the coups de grâce and Keith and I went over to pull both animals into the open where Pierce could lift and cut. The second steer kicked violently in his death spasm, narrowly missing Keith's head as he bent to hook a chain around its back hocks. "You son of a bitch," Keith said, looking at it closely to see if another shot was needed, but the animal was quite dead. "I've never had one do that before."

"Neither has he," I said, and we hooked the steer to the back of the pickup and towed him to the butcher.

chapter 20 • always, the grass

GARY WELCH AND HIS HELPER, LLOYD POE, WERE COMBINING barley at the Stevens, and I helped them in the morning, driving Gary's old green five-ton International, ferrying loads of barley back to Birch Creek, dumping them into the swirl of the big auger, up, up and into the top of the second grain bin.

Then just when I thought I'd done all the custodial work in sight, I was given the assignment of cleaning up the scale shed in preparation for shipping yearlings the next day. The scale shed wasn't used often, but it was cleaned much less frequently. Judging from the buildup of blown-in dirt, cobwebs and, over every surface, scattered scale slips like the ones I'd used in weighing hay, it had been a long time indeed.

Jerry came after a while to help, and I put him on window-cleaning duty while I mucked out the floor and tried to make order of the weighing tickets. After the Stevens, it was a piece of cake, a mere three hours. In the early evening we gathered yearlings, as gently as we could, so as not to run dollars off them; they would be weighed and shipped tomorrow. We parked them in the tank field, where they would be close to hand in the morning.

Bill would make money on these steers, but just how much would be hard to figure, after feeding them all winter, vaccinating them, getting them vital mineral supplements, and of course grazing them all spring and summer. After languishing in the 50-cent-per-pound region all year, prices were edging a few cents upward; still nothing to brag about. At those numbers, the profit on this deal wouldn't exactly leave Bill set for life.

Six A.M. found me at the corrals with Donnie Pettit, getting the horses in. Bill Loney was coming to help with shipping, and he and Bill would ride this morning as we moved the steers into the big round corral so they could be pushed through the alley, sorted, weighed and loaded onto trucks.

The horses roamed a large field to the south of the corrals. Sometimes they could be very difficult to catch in the mornings, but today at daybreak they followed Donnie and a bucket of oats across the creek and into the corrals nicely enough. We haltered Roscoe, Bose and Dusty, fed them oats and saddled all three of them just in case someone else wanted to ride, then kicked the other horses back out. "Better make a pot of coffee in the scale shed," Donnie said, and just then Keith drove up and told me to do the same thing.

Willie and Jerry showed up, and Keith put them to setting the gates so the steers could be run into the corral with a minimum of fuss. A few minutes later Bill Loney arrived, accepted a cup of coffee and even brought a tom turkey to join the Birch Creek crowd that loitered around the corrals like young toughs on a street corner, strutting and squabbling. The newcomer wasn't exactly welcomed when Loney opened the cage and booted him out; rather, a gaggle of half a dozen pecking and screeching hens sent him scurrying into the next corral, prompting the assembled cowboys to make a number of trenchant comparisons to human gender relations.

Bill arrived then, and Donnie, Bill, Keith, Willie and Bill Loney moved out to do the gathering, leaving Jerry and me back to watch for stragglers. Shortly before eight, a brand-new silver Pontiac coupe came whipping into the yard, suspension sorely tested by a couple of the larger potholes in front of Willie's cabin. Out hopped a flashy young man to match the car, wearing sprayed-on Wranglers and a grin that could only be described as cocky. "I'm Tim Sundling," he said. "Do you have another horse saddled?"

"Yep."

"Lead me to it."

The way he mounted up, I knew he'd been riding all his life, so I didn't worry too much when he rocketed Bose out of there as fast as he'd spurred the Pontiac in. He needn't have bothered; the crew had the steers under control, and in another fifteen minutes we were bringing them up the alley.

I got gate duty in the alley behind the scale, where Sundling, the cattle buyer, sorted with a decided flair, cutting back a few here and there, but not many. He and Bill made their deal in the scale shed, somewhere in the 60-cent range, and by nine we were loading the steers on the waiting trucks, including our two and one each from Ben and Errol. By noon we'd moved all eight hundred head.

It was considerably easier, all in all, than the way cattle were shipped when Bill first came to Birch Creek. The cattle were trailed to White Sulphur and loaded on Milwaukee Road train cars to be sent wherever the buyer wanted them. Often a hand would go along to tend them, and then when they arrived, they would have to be trailed again. Trucks hauling bullracks were simpler, but I thought it must have been fun, if somewhat cramped and pungent, for the hands who accompanied the stock on the train.

When we'd finished, Bill said to me, "Wear your going-to-town clothes tomorrow. You can ride along with me." He was driving the old yellow Kenworth to Billings with the cut-back yearling steers and some dry yearling heifers along with some of Ginger Kinsey's culled cows.

The next morning he called me at the bunkhouse at 6:45. "Let's see how handy you are," he said. "I want those yearlings moved across the alley into the loading pen. We'll put them on first. Then put Ginger's cows into the alley."

"What's so tough about that?"

"I'm getting to it. There are two pairs in with those cows. I want the pairs sorted off separately so they can be sold that way. They'll bring more. So you've got to figure out who the two mother cows are." I took Brent with me—Jerry asked Bill for the day off last night—and we easily got Ginger's cows into the alley. Then came the hard part. I thought Brent's dairy-farm experience might help, but when I pointed to the two calves in the bawling mass of cows and said, "Which two are the moms?" he just shrugged his shoulders and said, "Don't know. No way to tell. Ain't eatin'."

I watched them closely for a few minutes, but no maternal bonding was apparent. It was like one of those corny old English mysteries where the detective sits all the people in the mansion down in the drawing room and says, "One of you killed Lord Figginsbottom and nobody's leaving this room until we find out who."

"Let's sort the calves off and see if we can tell then," I said, and that got us halfway there. When we pulled the calves off, one cow stuck so close that I figured she was one of our two suspects. Another cow was sticking close to the fence and making a lot of noise, so I let her in, but she stayed away from the calves and kept mooing, so I figured she wasn't the one. Then we had to sort her off and get her back out the gate without losing the other three. All the cows were making some noise; getting sorted and moved was stressful and they'd had a rough day yesterday, getting shipped down from the Stevens. But finally I spotted one cow whose calls seemed different from the rest, lower and more strident, and one of the calves seemed to be answering her. Still, she didn't want to get sorted out of the herd, and it took us three passes through them before we got her. When we finally did, we were rewarded by the sight of her zipping right over to the odd calf, and in a few moments he was getting some breakfast.

Bill showed up just then, and surveyed the pens. "Looks okay. That one pair's right, but I'm not sure about the other one."

"Bet," I said.

"Well, I don't know, we'll see how they do."

So we loaded them up. The steers were a pretty sorry-looking lot. One little guy was far too small for his age; he'd probably been injured at birth. A couple of others had obvious flaws. But the one most interesting to me was a fat, healthy-looking black Angus yearling.

"What's the matter with him?" I said, pointing.

Bill looked at him closely, then chuckled. "I'll be goddamned. That's our waterbelly steer, the one I operated on. He sure bulked up nice. He might bring a pretty good price if nobody notices he's pissing like a heifer."

The ride was pleasant, the Kenworth humming right along on the two-lane beside the Musselshell River, eastward through Twodot, Harlowton, Ryegate and Lavina, turning southward there for the final forty-five miles or so into Billings.

Bill pulled into the sale yard and waited for an open dock; it looked like a pretty busy sale day. Then we unloaded the cattle and divided them into lots for sale. They were given a quick but practiced once-over by state brand inspectors—this was the job Willie John used to have—and assigned lot numbers. Bill moved the truck and we headed into the café for some lunch, and then into the sale barn.

The auctioneer, a young, dark man in a big gray Stetson, was keeping the lots moving through. "Look at these fancy young yearling bulls," he said. "This is Lot 213, average weight eleven forty-six. These bulls are going to help somebody's herd."

"They do look pretty good," Bill Galt said, eyeing the bulls. "Better than anything I expected to see here. I'd buy 'em and haul 'em home except you just don't know what you're getting, where they're from, what might be the matter with them."

"Will somebody buy them for stud?"

"Probably. They could just fatten them up and kill them, too."

They got hammered down for sixty-three cents. One of them took a pass at a cowboy in the ring, eliciting a titter from the stands as the man jumped to safety, then climbed back and helped herd them out. "All right, take a look at these calves. They've got a lot of growing to do and you might as well take 'em home and make the money. Nice-looking heifer calves, sixteen in this lot, No. 214."

Maybe sixty or seventy people were scattered around the bleachers, but three or four buyers seemed to be buying all the weight, including one lanky, long-haired character who looked like what

George Armstrong Custer might have looked like if he'd lived another twenty years. He was Rudy Stanko, a cattle buyer who was well known in Montana for his entrepreneurial efforts, including trying to start what would be the only slaughterhouse in the state; and also for several scrapes with the law and for his political views, which tended toward a far-right libertarianism tinged with white supremacy. Stanko recognized Bill and nodded in our direction. He was buying a lot of feeder cattle, and he was flipping through extensive notes on what trucks he had available to fill and what lots he'd looked at. We didn't stay to see our lots go through. Bill wanted to get back to the ranch. "I don't need to stick around and hear these prices," he said with a grim smile.

We were zipping along, making good time empty, just coming into Ryegate, when Bill said, "Oh, shit." He had flashing red lights in his rearview. As he pulled over, he said, "Call Ginger, will you, at this number?" He gave me the cellular phone she carried in her cruiser. I tried. "No answer, Bill," I said.

"Damn, she must be out of her car." We were on the side of the road, and up walked a stocky dark-haired officer with sunglasses, a big .45 and a star that said "Golden Valley County Sheriff."

Bill was already fishing in his wallet for his license. A ticket in a commercial vehicle was no laughing matter, and I knew he was concerned. "I was told to pull you over," the sheriff said, and Bill said, "Oh? By whom?"

"By me," said a voice at my elbow. Ginger Kinsey had opened the door on my side and was standing on the running board, grinning at Bill.

"This is going to cost you a cup of coffee," the sheriff said, grinning too. "By the way, I did get you at 71." Speed limit for trucks is 55. Ginger was laughing now, at Bill's red face. "Okay," Bill said, "I owe you one."

So we stopped in Ryegate and had coffee with the sheriff and Highway Patrol Officer Kinsey, who looked very squared away in her crisp brown uniform.

"I don't know what you got for your cows yet, but I know who probably bought them," Bill told her.

"Who?"

"Rudy Stanko."

Ginger laughed. "You're kidding. I just gave him a ticket the other day. Clocked him at 113."

Montana is still a small town, I thought with pleasure.

I spent the next day fencing stackyards, just me and the wind. When I got back to the corrals, I found Bill's pickup parked in front of the calving shed. As I passed it to go inside, I noticed the inside of the bed was covered with what looked like brick-red paint. It wasn't paint.

While I'd been fencing, the rest of the crew had been working Miller's forty head through the chute. Bill's border collie, Freckles, had been helping move the cattle. Somehow he'd gotten cornered by one of the cows and stepped on, suffering a very bad compound fracture of the hind leg.

They'd called Doc Schendal right away, but he'd been unable to stop the bleeding, and so he and Bill had bundled Freckles into the truck and then into the Cessna and flown him to an animal trauma center in Helena. Freckles had very nearly bled to death. The doctors in Helena had managed to get it stopped, but he was so weak they weren't sure he'd live the night.

Freckles was a favorite with everyone on the ranch. He had a wonderful, demonic grin, baring his teeth in pleasure and wagging his entire hindquarters when he saw Bill—particularly if he thought he might be in trouble. He was very well trained, as one would expect, given his ownership. As was usual with the best of his breed, he was extremely intelligent and a superior athlete.

"Go get a gopher," from his master, would turn him into a rodent-seeking missile, running at top speed across a field. Bill had even trained him to climb a corral gate—something I would not have believed if I hadn't seen it.

Bill's ingrained John Wayne stoicism would not let him show it most of the time, but I knew how attached he got to his animals. He'd recently lost an older collie, Doc, and he'd been distant and subdued for several days afterward. Tyson had told me Bill had buried Doc at the end of the runway behind his house.

I could tell that Keith was concerned for the boss, too, and the rest of the day was passed in the shop, somberly.

"How's Freckles?" I asked the next day.

"He's going to make it," Bill said. "We won't know about his leg for a couple of weeks, though. He'll probably lose it."

Right before Willie left, he sold his old Chevy pickup to Bill, who bought it pretty much as a favor to Willie, so he would have a down payment on a new truck. This morning Bill said, "Get that pickup of Willie's in the shop, get it serviced and install a radio in it. We might as well get some use out of it."

Last week, he'd also bought a three-quarter-ton Ford at auction, a former Montana Power vehicle, that seemed to be in very good shape. Bill had bought a flatbed earlier in the year that he'd planned on putting on Doc Schendal's truck. "Let's put that flatbed on the Montana Power truck," Bill told Keith. "That'll make a good outfit for Donnie to drive."

Brent got the flatbed installed, with some welding help from Keith, and also fixed the axle that broke on the four-wheeler. I got Willie's truck as roadworthy as I could. Call me Mr. Goodwrench, but I felt pretty good about finding a broken brake line during the service and replacing it, then bleeding the brake system. Then Brent and I installed radios in both trucks. A productive day in the shop.

Bill showed up in the afternoon, nodded approvingly at the progress we'd made, then asked me to go chase down a bull that had gotten loose and was on the road between his house and the corrals. "Just put him in the pasture in front of the corrals," Bill said.

I remembered Keith telling me that sometimes the gate by the cattle guard doesn't seem inviting to cattle out on the road and that they preferred the lower gate into the pasture. So I opened both and found the bull, a big, petulant black Angus who was about to go into the corner stackyard. I didn't want him in there tearing up fences, so I slipped over and got behind him, then blocked the entrance to the stackyard with my four-wheeler as he moved toward it. That sent him down the road at a run, which was okay with me. He was going the way I wanted.

"Okay, you big son of a bitch, door number one or door number two?" I said. As advertised, he disdained door No. 1, opting for a lurching right turn at the corner instead. So I got behind him a little closer and pinched him toward the open gate on the lower end of the corral, and the grass on the other side looked appetizing enough that he went through easily. I closed the gates and headed back to the shop. It had taken maybe four minutes, start to finish.

"What are you doing?" Bill said when I walked in. "Why aren't you getting that bull?"

"Because I got him already," I said. "He's where you wanted him." I tried not to look too pleased with myself.

"Wonders never cease."

Bill was building a road on Doc Schendal's place, through a section of second-growth timber that Schendal was about to log. Brent and I were tapped to help. I was frankly dreading it. Of course Doc has the right to log his private land, and of course Bill has the right to help. But building logging roads in forests is not something to which I aspire. I felt like a collaborator in a mugging.

"Go out to Schendal's. Go down the main road. You'll see the turn, and then you'll see the grader. Just follow the Cat tracks." Of course it was not nearly as simple as that, but after mistakenly driving onto the neighboring Hutterite colony, I finally found the right "main road," and shortly thereafter we were following the just-cut track through the trees.

The D-6 and the D-8 were both up there. When we arrived, Bill was working on smoothing a switchback with the D-6. Waiting for us were three huge sections of thirty-inch culvert and two collars—two fourteen-foot sections and one ten-footer, a total of thirty-eight feet of culvert. After considerable jockeying, we managed to load all of it, appropriately enough, on the back of the pickup that used to be Doc Schendal's.

"You're going to have to back up the hill and then down to the coulee where that goes. It's sticking too far out the back and the slope's too steep to drive forward," Bill says. "You need to get that pipe bedded for me. You've got to get it level with the existing grade, in line with the water as it runs down the bottom of the coulee. Once you've got it right, put the bands on, but make sure it's good and straight."

Brent said, "I'll drive." I willingly relinquished the wheel to him for the harrowing reverse run up and down the slope, and he accomplished it with élan. I got out and took a look around, and felt a whole lot better. This land was a typical second-growth forest of spindly uniform pines, most of them probably about eighty years old. The land was rocky and barren, and next to quite a bit of cleared land, including the Hutterite colony. I wondered how the hell they'd managed to log this the first time. With horses, undoubtedly, but judging from the size of some of the stumps it must have been quite an undertaking.

Pulling the pipe out of the truck was considerably easier than putting it in, and we rolled it roughly into place. Much of the bed was already reasonably smooth, but the last eight feet of the culvert was hanging out into space. It needed from one to three feet of fill under it, to keep it straight.

We attacked the good part first, pulling rocks out on high spots, smoothing, making sure it was just right. Then we struggled with the collars, which were every bit as much of a bitch as I remembered them being during the Keep Cool reservoir project.

Then we started putting rocks underneath the other end, rocks and more rocks. We put all the rocks we could find in the immediate area, and still we needed more. Brent moved upslope and started chucking rocks down for me to jam under the pipe, and of course the inevitable occurred—he tossed a grapefruit-sized rock down toward me, and it caught me on the shin and raised a lump of almost equal size.

Two hours of rock hunting and chucking and jamming gave the pipe a relatively stable bed. Bill showed up, took a look, nodded grudgingly, said, "Get out of the way," and began pushing fill on the culvert with the D-8.

We did a little tamping, removed some large rocks that could damage the culvert, and let the Cat do the work. When the culvert was covered, Bill gestured to me. "Hop up here."

Sitting astride the gear lever, I got a Cat-skinner's view of the power of the D-8, which I had to admit was quite exhilarating. It was like sitting astride an earthquake. The enormous blade moved an astonishing amount of whatever was in front of it—rock, dirt, branches—but it looked to be every bit as difficult to do as, say, fly a Cessna. Bill had a knack with machinery, but that didn't explain all of it. He'd obviously put in a lot of hours in this seat.

Right now a large stump precisely in the path of the road—shown by red surveyor's ribbons on each side—was giving him a lot of grief. He took several passes at it, and at one point, backed up against the stump, he wedged the blade against a standing tree angled in his path. "This spot here is how you kill yourself on this thing. Well, one of the ways."

"Thanks for the warning. I'll be going now. Would you leave me your new pickup?"

But he managed to extricate himself and me without anything dire occurring, and after another couple of tries, he levered the stump out, and we roared on across the side of the hill, taking the first cut. In a while we came to a clearing, and on the other side were three red ribbons tied to one branch. "You know what that means?" Bill asked. I shook my head, and he said, "That's the end of the road." He dozed a little farther, then said, "Okay, get off now. I'm going to smooth this a little and then we'll get out of here."

Brent and I walked out behind the Cat, which took forty minutes. The wind had come up, and by the time we got back to the pickup, it felt entirely too much like October. I didn't thaw out until I got under the warm dribble of the bunkhouse shower.

The next day, in the late afternoon, I put out salt in the Lone Section and in Little Birch. I could see why Bill enjoyed doing it so much, driving around and checking on the state of things, the weather, the season, the cows, the grass—always the grass. Salting the Lone Section, I thought about what a section of land like this would cost in Gallatin County, or even Park County, where I lived. Anywhere from a million on up.

How many head can you run on it? Fifty? Sixty at the most? Not year-round, you can't. You'd have to let the grass rest and grow. With cattle at sixty cents a pound, it wasn't hard to see it just didn't pencil out. And when you can't buy rangeland at a price that will allow you to run animals on it, the land loses its utility to all but the "gentleman rancher" or "sportsman," who couldn't care less about making money. Then you no longer have a ranch, but a privately owned theme park. To think of that happening here in Meagher County scares the hell out of me, but you can't be that provincial. Of course it's happening all over the West, and that scares the hell out of me for the future of ranching, of the way of life I was privileged enough to be experiencing.

Conservation easements, where land gets placed into an irrevocable trust that prohibits subdivision, are a solution that many ranchers find unacceptable. The easements may at least help keep the land in large blocks, pretty to look at, without trailer parks, but whether they will keep the land in the hands of working ranchers is another question. Fewer and fewer people are willing to make the lifestyle sacrifices that, say, Bill Galt makes, living the ranch twelve or fourteen hours a day, every day of the year, and of the ones who are, only a tiny handful can actually afford the luxury of making those sacrifices.

I couldn't solve all this, but as I put out my salt, I took a grateful look at the grass, and hoped that this place would look like this, and work like this, for as long as I could see it.

OCTOBER 15

Meagher County, Range Five, Township Nine

Crooked lines map water:
Butte Creek, Birch Creek, Smith River.
No hills on this map, no timberline, no winter.
Just boxes of dreams,
rectangles carefully annotated
by less adventurous clerical hands
with names of ones who dreamed:
Frank Powell, James Ingalls, Joseph Glen.
Henry Lee Farney.
Names of men who first said this is mine
and when they said it: July seventh,
eighteen ninety-eight, October eleventh,
nineteen hundred eighteen. No Gros Ventre
or Blackfeet ever said this is mine,
but these men did. Women, too:
Amelia Mayn, Mary Gillinan, Lillian Bing,
Lucy Folsom. Moose still
live in the willows along the crooked lines;
they have no regard for rectangles.
These names live in tattered Bibles,
weathered headstones, on this map
and in certain coulees
where horses paw through ice to drink:
Leo Havey, Albert Bruckert, Soren Volden,
William Totterdale.
Some proved up, some didn't, some died trying,
some headed back to St. Louis. Dreams
thinned out, rectangles got bigger.
A handful of great-grandsons hold here
with the stubbornness of junipers

rooted steep to hillsides.
Oakley Jackson's house still faces down the wind
in Section Twenty-six but less is left to show where
Salathiel Gurwell made his stand
on two scant quarter-sections. Would he favor
the gentled lands and ditches that curl across
the narrow bluff between creek and river
he claimed on the first of August,
eighteen eighty-three? Who did he love there?
What of the others? Lawrence Helfert,
John O. Hussey, James Devee?
Salathiel, a few of us know your ground.
We've fenced it, walked it, fished it, hayed it,
fenced it again when the moose broke through
in their regardless rectangular way. This is mine,
you said, and so it is.

chapter 21 • the gather

ON A CHILL SEPTEMBER DAY JUST AFTER HAYING WE TOOK THE four-wheelers into the Big Field to gather heifers and move them to pasture behind the corrals. There would be more cows to push out of here later, when the herd returned from Lingshire after weaning, but this was the first step in the momentous gathering and moving of cattle to prepare for winter. Bill must have felt the morning frost too, I thought.

I loved this work, scouring the country for cattle, getting behind them, bunching them up, moving them out. The heifers weren't that far away, fortunately, and it went pretty quickly.

Much of the area I covered was rough, sidehill terrain, with sagebrush hiding big rocks. Hit one of them with the undercarriage of the old blue Kawasaki I was riding, and I wouldn't have an engine left. I was careful, being mindful of Bill's admonition, repeated this afternoon: Don't tear up any more buggies.

Later, when Brent came to help, I relinquished my buggy to him and walked the rest of the way, pushing the young cows up the road on foot, just as I'd pushed them the other direction this spring.

The next morning Bill gave me an ass-chewing I didn't deserve. "What the hell were you thinking of, taking that four-wheeler up on the rocks last night?" he demanded in front of the rest of the crew in the shop.

I flashed a glance at Brent. I'd seen him take the buggy I'd relinquished to him up onto the red rock bluff between the shop and Keith's place last night, and had wondered if anybody else had seen it.

"Goddammit, I told you before you started out yesterday, keep those things off the rocks. Get off and walk if you have to go places like that. I saw you from the lead, and I asked Keith who was on the blue Kawasaki. Be more careful. And that goes for all of you." It was especially galling because I *was* walking at the time. But I thought of Pat Bergan, and Bill's new Honda perched at a crazy angle on the cliff above Thompson Creek, and stayed quiet.

The frost was getting thicker by the day. One morning about a week later, only one of the hounds, Angel, an older bitch who I'd really gotten to like, came out to greet me this morning. The rest stayed kenneled up, hunger losing the argument with the chill. It was odd, not knowing all their names; nobody who knew was ever around in the morning when I fed except Keith, and if he was here, he was busy. So I'd gotten to know most of them by their faces and personalities only. Angel had quite distinctively slanted amber eyes and a nice collection of freckles. She looked *rural* somehow, as only a hound can look, but smart, too. I was glad she was getting a solo shot at the food; she was evidently on the lowest rung in the canine caste system. I'd noticed she often got snarled and snapped at and didn't feed until the others were sated. Later I would find out that she was the mother to every other dog in the kennel. No way to treat Mom in her golden years.

I felt bad for these beasts, that they never got out except during cat season in the winter. They had a good-sized run, maybe fifty yards by twenty, but it had to be pretty boring.

The ice on their water was an inch thick that morning, and the thermometer by the shop and my plume of breath concurred: twenty-five degrees. And by nine the wind was blowing again, a lot colder today.

I didn't care. We were headed up to the Dry Range to gather cattle. We took two trailers, one of them filled with horses, the other four-wheelers. I was really hoping for a horseback day, but it looked as though during gathering, as during branding, the horses would be reserved for the senior hands. Today Donnie and I saddled horses for Bill and Donnie and Pat; the rest of us got four-wheelers, except for Tyson, who rode an old motorbike. We were to take this herd, about 350 pairs, down through Ellis Canyon to Rock Creek, where they'd have about a month's grazing before winter.

This was rough country, hilly, with enough lodgepole pine to make chasing cows a slalom event. Finally Jerry and I got thirty-five head parked in a big open park considerably upslope from Den Gulch. I kept them there while he reconnoitered. He returned in a few minutes, having found the right path down the mountain, and soon we had our charges joined with the main herd. Bill counted them through a gate as we pushed down the trail toward Ellis Canyon. We got a little sunshine then, but that was about the only improvement. These cattle did not want to go peaceably down the trail. They kept slabbing out around hills, down into ravines, always wanting to double back. When they're trying to turn back, the drag becomes very active. I was hopping off the four-wheeler constantly, chasing cattle on foot through thick forest, trying to get them lined out again.

Running through the brush after some miscreants, I nearly stepped on the biggest sage grouse I'd ever seen, so big at first I thought it was a turkey. Now there was a good dinner, but not only was I unarmed, I had two renegade pairs to get back around.

As the afternoon wore on I began to get a little better at anticipating the spots where they would make a break for it and cutting them off quickly. But it didn't always work. Keith, Willie, Tyson and Jerry were also constantly dashing through the timber. We had ourselves quite a rodeo.

Tyson's bike got a flat, so he took Jerry's four-wheeler. Pat Bergan wasn't with us, so Jerry took Bose, the big sorrel horse that had been saddled for Pat. Then Brent's four-wheeler went down, too, and apologetically he took mine. "That's all right, I've been on foot most of the time anyway, chasing these bastards," I said, and he laughed. Finally they seemed to settle a little. Jerry, on Bose, and me, on nothing, pushed them along at the drag.

As we passed through one particularly steep-sided area of canyon, I heard it: the bleat of a calf where none should be, in thick timber uphill to my right. I scanned the hillside and finally spotted a black calf in the trees, quite a way up.

"I'll go get him," I told Jerry, and he nodded. The timber was far too thick and the hill too steep for Bose.

This damned calf must be half mountain goat, I thought as I climbed at an angle, trying to get around him and drive him back down.

Each time he stopped and looked down, I'd get within a few feet of cutting him off. But he'd always see me and climb higher. He nearly fell once, and I thought maybe that would discourage him, but he righted himself and skittered on up the slope. And he was climbing a lot easier than I was.

After about fifteen minutes he went all the way over the top, probably a hundred and fifty feet above the road, and disappeared into the pines on the other side of the ridge. Just then I heard Bill's voice calling out, and looking down I saw him riding back down the trail with Jerry. Bill had come back from the lead, checking on our progress, and discovered Jerry by himself. He'd asked where I was, and when Jerry told him, Bill had apparently given him a good ass-chewing for leaving me on foot without a radio. I called out in reply, "I'm up here, Bill."

He peered up and caught sight of me, most of the way up the slope. "What the hell are you doing up there? Where's the calf?"

"He just went over the top."

"We'll be a month looking for him," he grumbled. "Get back down here."

I tried to descend with as much dignity as possible, and when I got back to the trail he said, "Jerry, you walk for a while. David, take Bose."

It had turned into a lovely, fragile fall afternoon, windless now, still surprisingly cool despite the sun. I started out behind Bill. Wrong. "Get up here next to me," he said. "This isn't a dude-ranch trail ride. You ride next to the boss, not strung out behind."

We walked the horses leisurely, talking and looking around the little canyon. "Be careful up where that dead cow is by the side of the trail," he told me. "A bear's been feeding on it and a horse'll smell that and try to blow up. If he does you've got to overhaul him, let him know you won't stand for that shit."

Bose didn't wait that long. Tyson had left the motorcycle on the trail where we could pick it up later, and when we got close to it, Bose went off in about four directions at once.

"Give him his head," Bill yelled, and in a few moments when Bose got all four feet on the ground at one time again, he said, "Now kick the shit out of him and make him go right up to it."

I did just that, but he was having none of it. As soon as he got within a few steps, he'd rear, and when I smacked him on the flank with the end of the reins, he bucked and kept backing up.

"Get off him and come take my horse," Bill said. "Watch." Bill got on Bose while I held his horse; he waited until Bose had calmed down, then walked him up toward the motorcycle again. It took three tries, but finally he got Bose to walk up and sniff the dreaded machine, then walk all the way around it without acting up. "Now watch him," he said to me as we exchanged mounts again. "Watch his ears, his head. Don't let him get away with anything. These ranch horses will kill you if they figure out you don't know what the hell you're doing."

I did a little better job of anticipating his discomfiture when we neared the carcass Bill had warned me of, and as soon as he'd even entertained the thought of going sideways, I got him lined out and he went past easy as could be.

"That's a lot better," Bill said. "You know, all this Horse Whisperer nonsense? It's the horse that does the whispering, and you'd damn well better be listening."

"Did you tell Robert Redford that?" Redford had been up a couple of weeks earlier, scouting locations for the film.

"No, I didn't want to discourage him too much. I knew that as soon as he saw me he'd want me to take over his role anyway, since I'm so much better looking."

"You're crazier than Bose ever thought about being." I knew, though, that Bill was the real thing with horses. I remembered the stories about him fixing horses for neighbors when he was a kid, and watching him I could see how it was possible. He *was* listening, sensing everything the horse was sensing, then guiding, reassuring, not verbally but with body movements. "Don't talk to them," he said. "Tell them everything you need to tell them with your hands and your legs and feet."

The trail wound back around to the mouth of the canyon, and when we reached a big clearing Bill stopped, so I did too.

"Nice afternoon, eh?" Bill dismounted and pulled a little camera out of his coat pocket. He photographed the canyon, then Bose and me.

"Get on that black horse of mine," he said. "That'll be an even better photo."

"Okay, but I don't know if your horse deserves to get his picture taken with me. Probably ruin him. He'll get a big head."

"Dusty will park you on *your* head if you fuck up."

As we turned to leave, Bill said, "Well, who do we have here?" Watching our photo session from about fifty yards away was a black calf, the one I'd chased over the ridge. He'd worked his way down the other side to this bowl and was on the way back to where he thought Mom was.

Bill radioed Donnie. "That lost calf is here in the clearing right above the canyon. Can you zip back up on a buggy and get it?"

"Sure, Bill."

A sweet September afternoon, a little saddle time, even a happy ending. Not too tough to take.

The gathering was progressing in earnest now, and I was getting an increasing amount of responsibility as Bill and Keith's trust in my judgment increased.

Bill pulled me off fencing detail one morning with a single key of his radio mike. "Drive up to the Stevens in the crew cab and hook up the big aluminum trailer. Fill it with four-wheelers and head up to the Clear Range road. I'll meet you there."

Pulling this trailer made me intensely nervous. It's big, and it cuts in really badly on corners. It's light when it's empty, and hard to control on the gravel. It's worth about $15,000, plus about $35,000 for the truck, and $50,000 that doesn't belong to me slewing strangely down these roads was stress-inducing.

Donnie and Bill were ahorseback, and they left Brent and Jerry and me to start pushing along the trail down from Den Gulch, yelling as we went to try to flush cattle out. We'd got about a dozen head when Bill called me on the radio. "I've got about a hundred head up here, and I need some help. When you get to the green tank, take a right and follow my tracks until you run into me."

I did as he said, hoping like hell I was on the right trail. There had been a couple of forks, and what I thought were his horse's tracks were very faint. In about fifteen minutes, though, I heard a lot of cattle, getting louder, and so I quickly got off the trail and behind a bluff, to avoid turning them back.

Bill nodded in approval when he saw me, then said, "There's three pair that split off back there. I can push these better with that four-wheeler. I want you to take Roscoe here, and go back and get them." As I scrambled off the four-wheeler and took his reins, he said, "Now, he won't stand for any wrong bullshit. You'd damn well better let him know you know what you're doing." I didn't get much of a chance to impress Roscoe. I went off at a trot in the direction Bill had indicated, and in about three minutes I'd spotted the missing pairs. Just then Donnie came right up behind them, pushing them along. I followed him back to the group Bill was pushing, and I regretfully traded seats with Bill again.

"Ah, well, that saddle's a little too big for me anyway," I said. "My ass isn't nearly as big as yours."

"That's why there was so much daylight between you and it, I suppose," Bill replied.

We met up first with the main group Donnie had gathered earlier, then with the stragglers Jerry and Brent had pushed up, and before long we had a stream of cattle maybe three hundred yards long, heading quite politely down the trail to the Clear Range.

For the next four hours Jerry, Brent and I played traffic cop at the drag, pushing them up, heading off the few who tried to double back, and keeping them from spreading out into the brush. We parked them for a little while in the first pasture we came to at the Stevens. Then we shuttled horse trailers and vehicles around, and Bill sent me down the county road to close all the gates for the drive to the barley field, just past the Halmes place, where they would stay until we were ready to bring them to Birch Creek for weaning.

Bill counted them out the gate—right at three hundred and fifty head, maybe missing two or three—and we took off on the last stretch. The cattle were tuckered out now, and at the drag I had the familiar job of pushing along the lamest and slowest of the herd. One old cow, we discovered, was actually blind, and once we figured that out, it was pretty easy to move her by sound and touch. The slowest, for a while, was a very late-arriving little calf. He was only a few days old—six months younger than the rest—and he was just about done by this time. I finally picked him up and put him on my lap and wrapped his legs around me.

Pat Bergan, heading for town, came up behind us in his pickup. "Could a fellow find work on this outfit?" he cracked at me, and I replied, "Sure, come carry this little shit for me," and he laughed and eased on by.

It was very nearly dark by the time we got them all in, and I was damned tired of carrying this little guy, who'd managed to kick me in the nuts half a dozen times in the last hundred yards, much to the amusement of Brent and Jerry.

Then Donnie yelled, "David, Jerry, Brent, come with me," and he took us up to the Stevens, where we turned the horses out in the corrals, gave them oats and put away the tack. Next we tackled the stackyard, where deer had knocked several panels down. We reset four panels in the dark, wiring them up to steel posts, and when we had finished that, we finally got to drive home. It was edging toward eleven when we got back to Birch Creek. "Be ready at six-thirty in the morning," Donnie said to Jerry and me. "We've got work to do."

"Fuck him," Donnie said at six forty-five, and started out of the yard without Jerry. We were almost around the corner when I saw him out of the rearview on my side, running out of the cookhouse, waving his arms.

"There he is, Donnie," I said. Donnie sighed, stopped and put it in reverse, and fishtailed back to where Jerry was standing, Stetson and chaps in hand.

"Didn't Brent tell you I'd be out in a few minutes?" Jerry said.

"Yeah, but didn't I tell you to be ready at six-thirty?" Donnie snapped. "You've never been on time yet."

Jerry had nothing to say to that. I got out and moved some gear around to make room for him, and he got into the middle between Donnie and me.

By the time we got to the highway, Jerry was asleep. Donnie slammed on the brakes. Jerry's eyes flew open. Donnie stuck a finger in his face. "Listen, you little shit. You're not going to sleep in my truck. If I'm awake, you're awake, understand? If you go to sleep again I'm going to hit you as hard as I can, right in the nuts."

"O*kay*. Jeez." Jerry lapsed into surly silence.

I thought about Donnie, sleeping in after the snowstorm at Hodson's, getting peppered with birdshot, and I figured he'd do exactly what he said. Jerry must have figured so, too, because he stayed awake. We were headed to Bill Loney's today, to help him gather a Forest Service section he leased in the high country overlooking his spread. We stopped at the Stevens to catch the horses, saddle and oat them, and by eight o'clock we were drinking coffee at Bill and Pam Loney's beautiful ranch house.

Pam, usually effervescent, was very quiet and sad today. Like her husband, she adored animals of all kinds, and one of her beloved cats was very ill. She was sending him to the vet this morning; she feared he would have to be put down.

Jill Stephens, a close family friend, had come over to help us gather cows, and before we went out she spent a few moments with Pam. Then the five of us mounted up and rode off to meet a beautiful morning, much chillier than the last few. It didn't take long to see that Jill, a pretty blonde mother of two from a prominent farming family near Dutton in north central Montana, was an excellent horsewoman. We all went upslope to get around the cattle. Donnie and I followed a creek bed up the mountain while Bill, Jill and Jerry circled eastward.

It was tough going, and when we finally broke out on top, we'd seen no cattle. So we walked the trail along a rock-strewn ridge, resting our mounts a little bit. It was still cold enough to elicit clouds of steam from John and from Donnie's horse, Susie, but the sun on my back felt terrific, and the smells of horse and leather and pine lingered sweet in the air.

We found a bunch of cows to chase through the timber. John, slowed by our relentless uphill climb, found his second wind, as he always did behind cows. He snorted and cantered down the mountain like a three-year-old, and we emerged with about twenty head. The other three had perhaps twice that many, but that left another large group of cows somewhere. Bill, Jill and Donnie circled for a second pass, and Jerry started downhill with the group we'd gathered. John was pretty tuckered, but I thought he could go a little longer, so we headed up to find a group Bill Loney had seen up a little draw near the crossfence. All but one were on the wrong side of the fence, and scattered when we found them, so John and I herded one cow back, then took the long trot down the fence to the gate we'd come out earlier and walked back up toward the others, John's breathing labored and rasping. It was rocky, dense lodgepole pine country, very tricky to maneuver, and John didn't like it one bit. But he hung in there gamely, and we got around three of the strays. As my legs brushed against tree after tree, I was grateful for the fine leather chaps my friend John Fryer had loaned me months ago. He'd get them back with a little less leather on them.

That last foray was as much as old John had to give. He was really played out, and I walked him slowly down to the creek for a drink, then on back to the trailer. He'd sure earned his keep today.

Bill was pleased with the count; only about ten head remained somewhere up on the mountain. "They'll come out," he said. "That's about what we get every year."

We came back and gave the horses some oats in Bill's barn. Loney came in and rubbed an old ewe named Eunice on the head; patted Joe, his old collie, who was Freckles' father; and said hello to a scraggly-looking rooster named Eddie. Then we went inside to a real bonus, something we didn't get often: a home-cooked meal, swiss steak, scalloped potatoes, corn, hot biscuits, the works. Two more cups of coffee and I figured I'd had my best day cowboying, no matter what happened during the last half of it.

We took the horses up to Lingshire, turned them out and tended to the tack. Donnie would use them over the next few days to start gathering the Lingshire herd, and then trailer them back to Birch Creek.

We headed back down to the Stevens, where more fencing awaited. I hadn't brought a change of footwear, and I discovered, as I'm sure many hands have before me, that cowboy boots are not exactly ideal when you're clambering over lava rock on steep slope. Fortunately, the fence Brent and I were going around was a decent shape, albeit perched on a rugged corner of the ranch. We pounded, stretched and clipped for a few hours, until the afternoon retreated into dusk. On the ride back to Birch Creek, I opened the window of the truck and let the autumn evening wash over me until Donnie growled about the chill, and wondered when I'd ever have a day to match the one I'd just lived.

"Jesus, that looks cold up there." Bill Galt was staring at The Needles, the rock spires high on the O'Conner, where we were headed to gather cows. I'd already made the same observation, which was why I was wearing long underwear, coveralls, two jackets and my winter wool Scotch cap.

Donnie and I got the horses, saddled and loaded them: John for me, Roscoe for Bill, Susie for Donnie, Bose for Bill Loney. It was much the same, for John and me, as gathering Loney's had been—a headlong, sidestepping skitter down pine duff and rocks—except that today the rocks had ice on them, there was no sun, and the wind was blowing hard.

And, of course, there were quite a few more cattle. We were busy from the first step, pushing cows out of the trees. We kept having to go sidehill to get around stragglers, and John wasn't the only one blowing hard by the time we hit flat ground a couple of hours later. John continued to work well as we emptied the cows into Little Birch, where they would stay for a day. I took him down into tough terrain along the creek bottom, hollering to roust cows from the brush. Each time we flushed some customers, John's ears would shoot forward and he'd hit another gear.

We got them all parked in Little Birch around five o'clock in the afternoon. The wind was increasing, if anything, and it had gone from chilly to plain cold. I helped Donnie with the horses and made a run for the bunkhouse, where I brewed up a pot of hot tea and chased cows down out of wooded hills and up out of brushy creek beds in my mind. I remember wondering, as I drifted into sleep, if John was doing the same thing.

"I think John deserves a day off today, don't you?" Donnie grinned at me the next morning as we got the horses and tried to decide who was going to get saddled up.

"He's earned his oats for the week," I said. "Who should I ride?" We were going to go bring the cows down from Little Birch, through a field called the Thousand Acres and into the tank field, where they'd be handy for weaning.

First, though, the Thousand Acres fence needed going over, so Brent and I had driven up through the tank field at eight o'clock this morning to take care of it, barely avoiding getting stuck a couple of times. And barely avoiding being blown into North Dakota. As Brent and I fixed elk-ravaged fence along the narrow bluff by the grand old white ruin of Oakley Jackson's homestead, the wind was so stiff in our faces that I found myself clinging to the fence like a lifeline. Not since the siding blew off the bunkhouse had I experienced wind like this. It had to be a sustained fifty miles an hour and gusts much higher.

By noon, though, we were done and down, the wind had moderated, and I was picking horses with Donnie, looking forward to another afternoon of cowboying.

"Hell, let's saddle Spider for you," Donnie said, "He's got quite a bit more get-up-and-go than John, and he's a damn nice old horse. He probably hasn't been ridden in at least a year."

"Okay," I said a little doubtfully. I had heard of Spider. He'd been Bill's personal horse for years. Spider was a big, handsome black horse with guts and style. The only problem with him was his fear of four-wheelers and motorcycles. He tended to blow up whenever they got near.

Now he was showing quite a bit of age. The hair on his face was slicker and blacker than the rest of his hair, and he'd thinned down a little, but he still looked strong. He took to the saddling good-naturedly, and as I led him into the trailer, Donnie said, "Be careful of him around four-wheelers."

"That's what I heard."

"He'll be all right, just keep a good hold of him."

Donnie was right: He had a lot more zip than John did. I rode him hard, and he seemed to love it. He was a dream with cattle. He had great instincts and he could cut quick and sure. Donnie took me up to the lead and showed me how to ride into the herd to break up a logjam and ride out again without disrupting the rest of the flow. "See there about fifty feet back from the lead, how they're milling around, not going forward?" Donnie said. "If you let that go, it'll get worse and everything will slow down. You need to cut in just below those cows that aren't moving, get them going and cut back outside before you get in the way."

I kept working them near the lead for an hour, moving in and out on trouble spots, making sure they were headed right. It was a lot more fun than cleaning up the drag, and a lot harder. I saw why the experienced hands always worked up here during a drive. Spidey had experience enough for both of us. He knew these moves cold, and if he was rusty, he sure didn't show it.

We did have a couple of tense moments with four-wheelers. He would hear one coming before I could, and when he did, I'd know it. His whole body would tense, and his rear end would slide out like a truck losing traction on a curve. I'd speak to him softly and leave him his head but let him feel my control of the reins.

As we passed the Jackson homestead, Doran Deal, who was pushing the drag, came up on one of the old Kawasakis to say hello. Spider gathered himself, trembling, ready to cut loose. I turned away from the buzzing four-wheeler and gave him some distance, and he settled down, although he didn't truly relax until it was well out of sight. I wondered if he'd had some bad experience as a young colt or if he just flat didn't like them. Either way, I couldn't blame him.

It was nearly dark by the time I oated him and turned him out. He flashed his tail and showed me his heels as he raced to join the other horses. Quite a bit of life in him yet.

When Bill counted the cows through into the tank field, he found out we still had some gathering to do, up on the O'Conner. We were short somewhere around forty head.

The only trouble was, Bill flew it early in the morning, and they weren't anywhere out in the open. They just about had to be clear up in the thick timber above where we gathered—timber far too dense for horses or four-wheelers.

The thermometer registered all of eight degrees when I fed the hounds, and it had warmed all the way up to twelve in the two hours since then. And to add some literal icing to the cake, it was snowing up there.

The crew was gathered in the shop, and we were discussing the situation. "Who're my good walkers?" Bill said, looking around. "Tyson. Who else?"

"Me," I said.

"You old fart, are you kidding?"

"No, you've got me worked into good shape. And I've always been a good walker."

Bill considered this. He looked at Jerry and Brent, neither of whom were doing any volunteering. "All right. You and Tyson. Keith, you'll have to take 'em in on a buggy as far as you can get. Brent and Jerry, you'll scour the meadows below the snow line on four-wheelers, pick 'em up when they come down. You walkers will wear orange vests so I can see you in the snow. I'm going to fly it, and I'll try to guide you in to any animals I can see." He squinted up at the mountain, cloud-shrouded. "We'd better get up there, the sooner the better."

Jerry had an orange mesh hunting vest in his car, and he loaned it to me, and Keith offered a pair of heavy lined leather mittens, which I gratefully accepted.

At the last minute, Doran Deal talked Bill and Keith into letting him come with us on foot. "All right," Keith said. "But you stay with those guys and don't do anything stupid."

The three of us each took radios; this was no time to be out of touch, and Bill would be able to talk to us from the plane.

Brent gave Tyson a ride up, following Keith. It was snowing hard, and it wasn't long before Keith's rear wheels were sliding despite an extra three hundred pounds of traction on his buggy, in the form of Doran and me.

When we were dropped off, Doran went with Tyson and I headed out by myself. I climbed up higher in the timber. It was snowing a little, and what was on the ground was about ten inches deep, unless you happened to hit a drift, which I did almost immediately, up to my waist.

As I clambered out I saw a flash of red Angus hide in the trees ahead, and I ran up and around. It was two head, young cows, and they started down toward a clearing, with me on their tail, then suddenly zagged right and fanned out upslope in the trees. Aha. I felt like a door-to-door missionary: They did not particularly want to be saved. This was not going to be easy.

"David, where are you?" It was Bill, on the radio. "Try to get out to a clearing where I can see you." So I fought my way downslope a little way to an open spot, keyed the mike and described my whereabouts. "Okay, I've got you, I think, yes." I looked up and saw the Cessna zip over, bank sharply right—typical Galt turn, I thought—and head around for another pass. "Okay, David, there's six head in the trees below you and to the east, under my right wing . . . now!"

I looked up, trying to gauge where his right wing was at the moment of "now," and slipped and fell on my face in the snow. No, this would not be anything like easy.

"Do you have them?" Bill demanded as I scrambled through the trees.

"No. No, I don't—yes! Yes! I'm behind them, but they're turning up! I'm staying with them. They're in sight, headed for the fence. I'm trying to get above them."

Tyson and Doran started over to help me and ran into eight head of their own, which they immediately took off chasing in the other direction. I was no match for these six by myself in the snow, I discovered. I spent forty-five minutes tracking them, circling around the face of the mountain. They'd go down for a while, but when we approached open ground they'd cut left or right and scatter back upslope.

"This weather's closing in. I'm not going to be able to stay in the air," Bill said. "You guys are on your own."

On my own was exactly how I felt, on my own and outgunned. The snow was falling harder. The ground was steep, the forest thick and cold and confusing. Racing along after hoofprints in the snow, I felt on the borderline of losing footing, of losing myself, of sliding down through rock and drift to fetch up broken-boned against some unforgiving deadfall, of rasping and gulping at the freezing air until all breath tumbled out of my chest, of my heart stopping somewhere below here in deep forest, the snow like a blanket to hide my failure. In my head I could hear Bill saying irritatedly to the medevac chopper, "He's under my left wing . . . *now!*"

I finally got four of the recalcitrant beasts along the fenceline and pushed them down to a green water tank on the edge of the open ground, but just then Tyson called me on the radio. "David, we've got ten head coming down the fence—*shit!* They crashed the fence and they're circling back up!"

"I'm right below you. I'm coming up to help." I left my four by the green tank, like a teacher depositing troublemakers in the principal's office, wishing never to see them again. I had the remote hope that they would either keep moving lower and get picked off by Brent, Jerry, Keith or Donnie, or at least stay near the tank so we could pick them up when we brought the others down. Then I took another deep breath and charged upslope like a wounded moose. I knew Tyson and Doran were moving fast uphill after cattle and they had a head start. Not to mention having legs that were twenty-three and thirty-two years younger than mine.

I found the spot where they'd ripped through the fence and went that way. Lots of hoofprints now, along with bootprints in two sizes. I caught no glimpse of the cattle, but after about ten minutes, I did reach Ty and Doran. They'd stopped in frustration, trying to decide what to do next, and I bent over, hands on knees, and gasped like a beached fish for a minute or two, then stumbled over to a fallen tree and sat down.

"The dumb bitches," Tyson said with a great deal of feeling. "We ought to leave them up here and let them freeze." Instead, we trudged uphill to a gate crossing and found them, on the wrong side of the wire, looking as exasperated as we were. "I've got an idea," Tyson said. "Doran, you slip through the gate and hide behind that tree. We'll watch and if you don't spook them, we'll get around and bring them down this way. Stay hidden unless they don't go through the gate, then try to head them off."

Doran secreted himself successfully, and Tyson and I curled in a great exaggerated circle around the cows, approaching within about sixty yards straight above them, and crouched below a little mound, watching. "Okay, now," Tyson said, and we stood up and started moving.

The cows moved agitatedly toward the gate. So far, so good. The leader—a red Angus I suspected was the one I'd been chasing all morning—stopped by the gate and looked around.

"Not yet, Doran," I yelled. There was still room for her to veer to the right, away from Doran and the gate. But too late. Doran jumped out from behind the tree.

Fortunately it worked. The cow nearly jumped out of her skin, but when she bolted, she went right through the gate, and the others all followed. Doran had saved the day.

"Yes!" Doran pumped a fist in the air triumphantly, and Tyson and I high-fived him as we ran past. "Come on now, Doran, we've got to get them down to the tank and the road before they decide to turn around," Tyson said. Doran closed the gate and we spread out, trying to keep them from turning back.

Like a snowball rolling downhill, they picked up more cows as they went, the red Angus clearly still in front. They were moving so fast I didn't see how they could possibly keep their footing on the slick hoofpacked snow and ice. We got them to the tank, maybe twenty-five head, streaming down the road now, lined out and moving. I thought then that we had them.

But at the last possible second before passing the last of the thick pines, the red cow lurched to the left, and then the whole smooth line disintegrated and spread out into the timber, melting from the road like a mirage.

"Fuck!" Tyson shouted. "Quick, after them."

We dashed into the trees. I saw Doran slip and cartwheel spectacularly over a downed tree into a snowdrift. He jumped up, swearing fluently but unhurt, and we plunged on.

One of the trying things about this sort of a chase is that cows are about two feet shorter than cowboys. Therefore they can move freely in dense forest under low branches that force us to duck and scrape and bang and swear.

They were losing a little steam, though. It had been an active morning for them, as well, and this last run had taken quite a bit of the whoopee out of them. In another twenty minutes, running along a high ridge, we caught up to them, and radioed Bill and Keith to tell them where we were.

"I've got to be close to you," Keith responded. He was up above us on a bluff, glassing the hillsides, looking for the cows and for us. "There they are, the bitches," he said after a while. "I'll be there in a few minutes." He came flying downhill, the high-rev snarl of his dirt bike a most welcome sound. "Bill, they've got thirty-seven head," he radioed a few minutes later. I must have undercounted them, or they'd picked up a few more on their last dash.

"Excellent. Keith, can you get them lined out from there?"

"Yep, we've got 'em," he said, and then he was there next to us, and we had them all in open country, where we could push them the way we wanted.

Brent came up to us in a few minutes on the red Kawasaki, and Doran, who looked more tired than I'd ever seen him and very white, climbed on the front. "I don't think I can take you, David, you're too big," Brent said. "Tyson, hop on."

So I stumbled on, my head pounding now from the glare of the snow, walking behind them until we'd gotten nearly to the stackyard by the O'Conner buildings. Jerry picked me up, then, on a four-wheeler and delivered me to the stackyard, where Bill was waiting with Roscoe beside his pickup.

"You look like hell," he said. "Bring my truck." I drove his pickup behind the drag, and it was one of the hardest things I'd done all day because my legs were so tired that every time I had to hit the gas, use the brake or the clutch I'd get a hell of a charley horse—first in my calf, then in my quadriceps. Finally when we were finished I was able to get out and walk around until my legs quit cramping. The feeling of completing this day was the sweetest yet, because it had come at a price higher than I'd known I could pay.

OCTOBER 20

Ain't This Something

I want you to gather the Dry Gulch today.
Saddle Spidey. Make sure
you check the brush down Thompson Creek.
Make some noise in there. Yell. Be careful. Niggerheads
and badger holes break legs. Cows', horses', yours.
Donnie saw a patch of spurge down there
somewhere. Find it. Remember Spider
hates four-wheelers. He'll kill you to get away from one.
Hell, boss, I hate 'em too. When did my life sell out
sidehill down a ravine? When did the trail
peter out to nothing? Easy. Just before
it widens out into the big green valley where the
creek runs with Walker's bourbon
and the cows gather themselves. The horses
all love four-wheelers because they make no noise.
If you could get there you'd never
get stuck to the hubs in mud
or bucked off onto shale rock. No spurge,
just fat cows and ice-cream grass. Your kids
are playing there, waiting for you without regard
to custody arrangements, and the magic badger holes
don't break anything. No frost heaves
because there's no frost and anyway
nobody would ever call 'em niggerheads.
Your grandfather is fishing in the burbling
bourbon. "Ain't this something," he says,
landing another whiskey-drunk brown trout.

chapter 22 • the end of the trail

THE HUNTERS ARE HERE, THE SPORTS, THE FELLOWS WHO PAY thousands for a chance, or many chances, at a trophy elk and anything else they can get a tag for—mulie, whitetail, bucks, does.

This is not something I do. I do not oppose hunting on moral grounds, although I have a little trouble with the concept of trophy hunting as opposed to hunting for food. I love reading the masterful writers on the subject, from José Ortega y Gasset to Jim Harrison to Jim Fergus to Bob Jones to Guy de la Valdene. I do bird-hunt, infrequently and badly. And I certainly love to cook and eat elk, venison, and antelope. But I never have gotten any kind of a yen for shooting a large animal.

Not so these guys. They are here, and they are excited. The ranch has a great elk herd—greater than Bill would like, frankly, considering the damage they do and the grass they eat. They are not natives; the elk here are descendants of a herd planted by the Fish, Wildlife and Parks Department in the 1940s near Canyon Ferry Reservoir. And they have thrived. Bill estimates he has well upward of a thousand head, just around Birch Creek, and many more in the Lingshire country.

Keith relishes hunting season; it is a break in the work routine for him, he enjoys the hunters, and he makes damned good tips. If anybody knows where animals are on this ground, it's Keith. Bill and Gary Welch help out, too, as well as hired guides Andy Celander and Aaron Florian.

My participation is limited to the grittier end of things, wrestling animals into and out of pickup beds, helping chain and hang them, washing blood out of the trucks before it freezes, chainsawing through skulls and brains to remove mounts, taking trimmed-out carcasses and barrels full of frozen heads to the dump. None of this bothers me, any more than the other time spent cleaning and doing dump duty; it has to be done. But neither does it engender any change of heart on my part. Bill has an annual hunting day for the crew, at the end of the season, when everyone can go up and blast away at cow elk. I don't think so.

Now, in late October, there was no getting around it: Winter was here again. The shop thermometer hovered at zero and the wind was blowing hard.

Just before dawn a truckload of corn arrived from Nebraska. I walked out of the trailer, helped Brent jam the auger under the trailer's first trap, started it running, opened the trap and gave my back a hell of a workout.

We climbed through a mountain of corn to get to the top and then shoveled like hell, no resting allowed, until the trap was empty, then moved up to the front of the truck and did it again. This compartment was bigger because of the nose. We tossed our shovels up through a tiny opening and then crawled after them. "Come on, hurry up, let's get it done," Keith exhorted, shoveling beside me and moving a hell of a lot quicker than I was by now. Shoveling corn is nasty work, and my back suddenly felt like it did last winter. Not a nostalgic feeling at all.

I went down to the corrals then with Brent and ran eight head of sick calves through the chute, vaccinated them, and took them back into their respective fields—heifers below the road, steers in the steer pasture like last year.

Then Keith, Brent and I got on four-wheelers and absolutely fucking froze, taking fifty-three head of old cows from the field behind the corrals up to the tank field. My hands and face were much colder from the four-wheeler in the wind than they'd been all last winter. It was slow work, pushing these old gummers up the hill. They'd go only so fast, staggering, crippled, but we'd gotten them into the field and most of the way back to the shop before Keith talked to Bill on the radio and found that six head were supposed to be held back for some reason. The wind was blowing snow, whether off the hillsides or current snowfall I couldn't tell. My face felt like it was falling off. Keith was yelling at me, what I couldn't tell. He was turning the cattle, then screaming at me to push them, then to back off. Turned out he'd seen the ones he needed, and he got them cut out, steam plume billowing from his mouth as he swore, and finally we got the rest through the gate again and pushed the others, back through the bottoms to the corrals.

I tossed them a little hay on Keith's instruction, and then Keith came running out of the calving shed, yelling, "Stop feeding them. Bill wants them trucked to the River Field."

So we loaded up, took them out to the River Field, dumped them out and came back to the corrals, where we found twenty-four tons of salt on pallets that needed to be moved into the two reefers very quickly so the snow didn't get it wet. It does not stop.

Stowing the salt was a tricky two-hour process, Keith pushing them up with the loader forks and Tyson, Brent, Jerry and I maneuvering them into place with the hand pallet jack. My wet boots slipped constantly on the metal floor of the reefer. We were working in tight quarters, and the pallets were very heavy and hard to move. Before we could get them put away right we had to empty the reefers of all the bands, broken pallets and cardboard sleeves from the old shipment of salt. We stacked all of that stuff up by the trash barrels and burned it, and I was grateful for the warmth.

We ate quickly then and went back to the shop, where I serviced Keith's truck. He went out in the afternoon, hunting, and when they returned after dark he'd gotten his charges well on the way to filling up their tags. Keith spotted a nice mulie in the brush near the creek side stackyard on the Gurwell, and his hunter Gary got it. Jay, perhaps the least experienced and skilled of the hunters, shot a nice five-point elk near the O'Conner buildings, and he was in that manic post-kill euphoria, describing the scene and the kill to all.

We hung the mulie on the tractor that was hooked up to the auger, and I wondered how that might discourage the deer I saw early each morning on that very spot, looking for spilled grain.

Another returning hunter bragged, "I shot this little doe right in the eye at 250 yards, see?" He held her head up for us to admire. More brain-salad surgery with the chain saw. Enough, I think, to last me several hunting seasons.

Just before six A.M. on Halloween morning, as I was pouring my second cup of coffee and buttering toast, somebody hopped up the bunkhouse steps. Like an old man set in his ways, used to being alone at this hour, I was a little disconcerted. It turned out to be Bill Galt, come to give me work.

"Keith and I are going to be busy with the hunters this morning, and Donnie's up at Lingshire. After you feed the dogs, can you bring in the horses?"

I'd never caught the herd of horses by myself before, and they were sure to be skittish in the cold, and I couldn't wait to try. "Sure," I said, not sounding it.

"Just go out with a bucket of oats and get them, or run 'em in with a four-wheeler if you want. Then take, let's see, the new maroon flatbed and hook up the white trailer to it, then load it with four-wheelers. Gas them up and get them ready. Oat the horses and leave them in the corral. I have to fly Freckles to Helena for his leg. When I get back we'll go to Lingshire and start gathering cows."

I got my bucket of oats and looked up at the horse pasture. They were nowhere to be seen. I walked up a ridge, across a frozen creek, then up another hill, and there they were, in a little knot, enjoying the morning together like a bunch of coffee drinkers, quite interested in me but not in the least alarmed.

John and Roscoe and a bay horse I didn't know came first, then the rest, shouldering each other, trying to reach their long snorting noses into the bucket. I turned and walked down the hill and they followed. They had a few moves to try on me. First Roscoe, then John bumped the bucket hard from behind, like defensive backs trying to knock the football loose from a ball carrier. Then Susie bumped me in the back a couple of times, even nosed my arm, but I knew that bucket was my only weapon and I held tight to it.

They didn't want to cross the frozen creek. I crossed it twice myself, showing them it was fine, but they sniffed it doubtfully and held their ground on the other side. I was equally obdurate, holding the bucket out temptingly.

Finally Spidey and Roscoe led them across, and from there it was easy. They went right into the corral, and I closed the gate, got another bucket and fed them all oats while I hooked up the trailer.

When Bill returned he told me to saddle the usual suspects—Dusty, Roscoe, Susie and John—and I did so, relishing it, taking my time. I gave them each a little more grain, curried them out carefully, put some red disinfectant on a little cut on one of John's hocks, saddled them up, loaded them and turned the rest out.

After all that, I didn't know if I'd get to ride, but I did, Bill and Loney and Donnie and me. Donnie and Jerry and Pat Bergan had already got the herd most of the way in, and we moved them close enough to get to easily when we started weaning tomorrow. It was another beautiful afternoon spent in the company of a few hundred cows and calves, with a horse I admired under me and good cowboys on either side. Everything happened the way it was supposed to; it would have without me there, but I felt, as I had lately, that I'd been useful, and there aren't many better feelings.

Keith had met us up there, and on the way down the hill toward the Stevens he said, "Did you two guys get those three head the other day?"

"No, Keith, we couldn't," I said, and told him where the three outlaws had gone and how they behaved.

"Well, let's get them now."

We unloaded three four-wheelers, and Keith, Brent and I went hare-assing across the clear range and found the two bulls and the delinquent heifer right where we'd last seen them. The next twenty minutes were the most hair-raising I'd ever spent on a four-wheeler. They didn't want to be caught and Keith wasn't about to let them go. I made one ditch crossing airborne, sideways, landed badly on a little hillock and came as close to rolling at speed as I'd care to. A couple of low branch-whips across the face, a mad chase up a sagebrush hill, and we had them cornered in a little draw and out of breath.

The only problem was, we were a hell of a long way from anywhere and it was very late in the day. So we took them straight up and over a big ridge, through a gate and down to the county road above the Stevens. Now we needed to duplicate the last part of the cattle drive the other day, to the barley field, and dump these guys in there.

The heifer and one of the bulls were fine, but the biggest bull, an old horned Hereford, kept trying to go back, feinting charges at the four-wheelers. A bull threatening a charge is no laughing matter. On a four-wheeler you have no protection; you're pretty much a sitting duck, and we gave him a lot of room. He was pretty tired from our mad dash. His tongue was out and his mouth was dripping with foam. We cajoled him up the road. He'd take a few steps, then stop and glare at us for a while, then take a few more. Finally, just past Halmes' house, he stopped altogether in the ditch by the road, glaring, swinging his big head from side to side at Brent and me. "I'll take these other two up and get 'em in," Keith said. "Stay here and watch out. He's really pissed."

When Keith got back, he hadn't budged. It was nearly nine P.M. Keith finally said, "Fuck it, leave him. He won't go far tonight. Pat'll come down in the morning and get him." So we loaded the four-wheelers and drove past him as he stood in the road, panting, the murderous look still in his eye as we caught him in the sweep of our headlights, this horned, red-eyed, scraggly orange-haired beast, foaming at the mouth, left to his own devices on Halloween night.

* * *

The next morning Bill Galt and I drove to Lingshire, followed by the rest of the crew. Bill Loney and Pat Bergan and Tom O'Donnell from Avalanche Ranch had the coffeepot on the stove at the ranch house by the time we got there. Jill Stephens came to help, and before too long Doc Schendal drove up too. It was weaning and shipping day for the main herd, and that was a good day.

I worked the brush by Rock Creek, sending a few pairs up onto the flat toward Keith and Brent, scrambling back and forth across the creek, getting a refreshing amount of cold water in my boots. It didn't take too long to get them into the corrals, but once they were in there, oh, my. I'd never seen a bigger bunch of squalling, shuffling cows and calves, twelve hundred plus pairs, straining the stout old corrals so that I was amazed the rails didn't pop off the posts.

O'Donnell would let maybe fifty pairs at a time into the round corral, and he and Donnie and Keith and Bill Loney would sort the calves off into one pen, then send the cows through to Bill Galt and me. Bill was sorting culls, and it was my job to push them up in the square pen as fast as he wanted them, no faster and certainly no slower.

When the corral was full it was absolutely impossible not to be stepped on or kicked. Bill went down once when a cow caught him right in the knee, and I was afraid he was badly hurt. His face was white with pain, but he got up and swore and limped and kept sorting. Somehow the calves had a knack for kicking you in the same spot on the same knee where the last one had kicked you.

Once we'd knocked off all the calves, we had to separate them, heifers and steers, and that's where the split ears on the heifers really helped. By noon we had half of the calves weaned, and Julia Short arrived with lunch, just as good as it had been at the brandings. It was a great pleasure to sit on the floor in the old dining room of the ranch house, leaning against a bunk, eating a second helping of short ribs and potatoes and salad and Julia's frozen pink stuff, and listening to Donnie Pettit reminiscing with Bill Loney about Hodson's, and Bill and Keith talking quietly about the hunters, and the winter to come, and Jill and Julia chatting about mutual friends in White Sulphur, Doc Schendal's braying laugh leaking into the room from where he sat in the kitchen.

After lunch we started loading trucks with calves, and Bill and Keith took the last two, a little after six. By that time, in between loading trucks, Brent and Jerry and I were helping Loney and O'Donnell and Donnie move the cows up for Doc Schendal to preg-test. Jerry and Brent left with the last trucks, and I was assigned to stay overnight and help with the rest of the preg-testing in the morning.

I carved my name on the wall of the barn, next to Pat Bergan's, that night. Then I cooked dinner, leftovers from lunch mostly, and beat the tar out of Tom O'Donnell at cribbage, and drank the last of that ancient Johnny Walker I'd noticed up here in the spring. I reminded myself to bring another bottle up to replace it for next year.

Next year was a concept I didn't want to think about tonight. Instead I just looked forward to tomorrow, which would be my last on the ranch, at least for now. I'd get up, cook breakfast for Tom and Donnie and make sure there was coffee on when Loney and Doc Schendal arrived, and we'd move the other six hundred cows through. And that would be it.

After weaning, the cows' cycle, and my cycle, was done. Ahead stretched the winter: calving out the heifers, feeding hay, doctoring sick calves, breaking the ice for the bulls to drink, fishing newborn calves out of the creek up on the O'Conner and trying to save them. Then trailing the older cows back up this way in the spring. Fencing. Branding. Irrigating. Haying. Gathering. Weaning and shipping again. This year I had been allowed to be part of Birch Creek, and it had become part of me forever.

I hoped my fences would hold for a few years, and the hay I stacked would nourish some of Bill's cows through the winter. I hoped Freckles and Pam Loney's cat both got better, and I hoped Keith and Donnie wouldn't get tired and quit, and that Wylie and Tyson stuck around another year, and that Bill wouldn't fly the Cessna in bad weather. I hoped Willie John did well in the new cowboying job I'd heard he got, and I hoped Jerry got along okay, and Brent decided to stay the winter. I hoped Fletch and Christian were doing well. I hoped somebody remembered to feed the hounds tomorrow morning, and I hoped the swather wouldn't hit any rocks on the wheel line next summer. I hoped a few of the heifer calves I'd pulled ended up in the herd for years to come. I hoped Spidey and John wintered okay, and I hoped Pat Bergan got that damned Halloween bull tucked away somewhere. I hoped Super Chicken was still out there, and that the PTO wouldn't stick on the diesel all winter. I hoped nothing would change this place too much. I hoped that when I came back Keith would be grinning, the maroon truck would be running perfectly and there wouldn't be too many jugs to clean. And I hoped I'd be there again, some year, on the day when Bill Galt could look around at the new grass and the new calves and say, just like Ian Tyson: *made it through another on the northern range.*

acknowledgments

I will always be grateful to Bill Galt for hiring me, for helping me, for his friendship. I was very fortunate to work with all the hands I have mentioned. Donnie Pettit, Tyson Hill, Keith Deal and Willie John Bernhardt, in particular, were fabulous, and I am fortunate to count all of them as friends. Thanks are also due to my friend Sid Gustafson; his mother and father, Pat Galt Gustafson and Rib Gustafson; Bill and Pam Loney; Jack Galt and Louise Rankin Galt; Ben and Errol Galt; Gary Welch; Rose Pettit; Verl Rademacher at the *Meagher County News*; Milla Cummins, our wonderful Park County librarian, and Deborah Giron, Meagher County librarian; Ron Burns and George Zieg; Park County weed-control officer Clyde Williams; Lee Rostad for her historical assistance; and John Fryer, for his friendship, his wisdom and his chaps.

Suzanne Gluck at ICM continues to be the best agent a writer could ever hope to have. Thanks, too, to Lou Aronica at Avon, for his faith and vision; and to Patricia Lande Grader, who was both patient and correct, and therefore a terrific editor. Also, Karen Gerwin-Stoopack and Laura Richardson, for their many kindnesses.

For inspiration and support, I thank Ralph Beer, Peter Bowen, Tim Cahill, Russell Chatham, Jim Crumley, Martha Elizabeth, Jim Harrison, William Hjortsberg, John Holt, Jon A. Jackson, Greg Keeler, Bill Kittredge, Peter Matthiessen, Scott McMillion, John McPhee, Hunter Thompson and Paul Zarzyski. Also Ian Tyson, Chris Wall, Guy Clark, Wylie Gustafson, and, as always, Jerry Jeff Walker.

Every Montanan should be grateful for Richard Hugo, and for Joseph Kinsey Howard. I am.

Thanks to Greg and Penny Strong.

Most of all, thank you, Sarah, for everything.

about the author

David McCumber is an award-winning journalist, a former assistant managing editor at the *San Francisco Examiner*, and the founding editor and publisher of *Big Sky Journal*. He has worked for more than twenty years as a writer and editor at newspapers and magazines across the American West. McCumber lives in Livingston, Montana, but is currently on the road working on his next book.